T0213281

Audio Technology, Music, and Media

Julian Ashbourn

Audio Technology, Music, and Media

From Sound Wave to Reproduction

 Springer

Julian Ashbourn
Verus Mundus
Berkhamsted, UK

ISBN 978-3-030-62431-6   ISBN 978-3-030-62429-3   (eBook)
https://doi.org/10.1007/978-3-030-62429-3

This Springer imprint is published by the registered company Springer Nature Switzerland AG
The registered company address is: Gewerbestrasse 11, 6330 Cham, Switzerland

Foreword

Music is part of humanity. It has been so since Neolithic times, if not before. We know from their wall paintings that the ancient Egyptians loved to sing and dance and had some surprisingly sophisticated instruments. And so it has been for every culture ever since, music reflecting the inner spirit, the attitudes and feelings of people at any given time.

Consequently, we have a wonderfully rich heritage of musical expression which, with suitably skilled performers, we can recreate today and enjoy, just as it was conceived and enjoyed at its origins. However, it is important that musicians take care of every single detail in the music and perform it, just as the composer intended. Some of the greatest composers, such as Beethoven, were quite particular about how a piece should be performed and would often write instructions in the margins of the manuscript. We must then, surely, follow these instructions as honest servants of the music itself, as bringers of a noble message, in order to bring forth the essence, the beauty and dynamics of it.

It follows then, that, if we record the performance, this too must be undertaken carefully and should provide an accurate record of the event. The orchestra should sound like an orchestra, with everything placed accurately in space and every instrument heard at the correct level. At the rehearsal, the conductor goes to enormous lengths to get the sound just right at the podium position. This is the sound that those in the first few rows of the audience hear, and it is this which must be captured within a recording.

I have known and worked with Julian Ashbourn for a few years now. He strives for perfection and reaches it through his recordings. Julian has recorded every performance of the Hemel Symphony Orchestra, which I have been honoured to use for reference purposes after the event. His deep knowledge of both technology and music is extensive and it is with great pleasure that I see he is passing this on for the benefit of others. I have no doubt that this book will be highly valued by many in the music industry, as it will be by me.

Claudio Di Meo
The Kensington Philharmonic Orchestra
The Hemel Symphony Orchestra
The Lumina Choir
London, UK

Preface

Audio Technology, Music, and Media is an invaluable guide for all who love music and who are passionate about capturing musical performance. The book serves both as a fascinating history of recorded sound and a valuable handbook for audio engineers who have grown up with digital audio and computers and may have missed the fundamental skills which remain essential to obtaining good results. In particular, the use of microphones and the understanding of the space in which a performance is taking place. In the 1950s we understood this as we had been feeling our way with tape recorders and moving coil microphones since shortly after the war. The established record companies developed their own techniques, often based on the pioneering work of Alan Blumlein who was sadly lost to us during the war, but who had defined stereophonic sound and moving coil microphones as early as 1929. Companies such as Decca and EMI in particular made some wonderful recordings in these early years, helped by some excellent producers and conductors. Many of these skills are rarely used today, for various reasons and generations have grown up without an understanding of stereo sound or even high-quality sound.

There were several milestones along the way of this development and these are covered in some detail within the book. Firstly, the Second World War was pivotal in driving the development of technology in general and one example of this was the development of the V-Disc for US troops and the use of different materials for the records themselves. A more significant development however was that of the tape recorder which, while it was invented before the war, saw rapid development as the capacity to record speeches and events and broadcast them at a later date, or multiple times, was quickly appreciated. This was also to drive the post-war development of stereophonic sound, which now became a practical possibility. Developments in both broadcast audio and recorded sound took place in the immediate post-war years and led to the development of high-fidelity sound for reproduction in the home, and a huge associated market. Developments in electronics also helped Japan to rebuild its economy and all of this is covered in the book.

There were subsequent milestones in the development of purpose-built recording studios and a wealth of associated equipment. In the popular music field, this reached a peak with the 24-track tape recorder and some truly huge mixing desks, together with a plethora of external effects that could be incorporated into the overall mix. This would be seen by some as both a blessing and a curse as it encouraged experimentation but discouraged an orderly approach to making music. Such developments also had an impact in the classical music field as producers and audio engineers explored the possibilities offered by 24-track recording. This led to a sidelining of true stereo sound, to be replaced with multiple monophonic sounds positioned between left and right artificially with an electronic pan control. This also destroyed the natural sound of both the orchestra and the venue, but provided a clean, crisp sound and the ability to make endless edits in the performance. Such developments are covered and explained within the book.

Consequently, factors such as microphone technique and true stereo sound are covered in some depth, before moving on to the digital domain and what is important to understand in this context, as there is much more than simply high-resolution audio. The digital world is thus explored, including the importance of analogue to digital and digital to analogue convertors and much more. Naturally, given the revolution in the playback of audio with digital devices, loss-less compression algorithms are explored as well as recording with digital devices. Indeed, throughout the book, there is a good deal of instructive detail for the audio engineer, from the alignment of tape recorders to the use of compressors within digital recording devices.

There has been a parallel development in musicians themselves, both those playing the classical repertoire and those exploring new musical forms. This is obvious when comparing the recordings of classical works separated by 50 years or more. With popular music, the music itself has changed beyond recognition, as has the way in which it is produced. The advent of the Digital Audio Workstation has been a significant factor in this respect and, consequently, this is covered in some detail. This brings into question whether changes or advances in technology are necessarily good. Within the musical arts, one might posit that it is the performance itself which is most important. In this respect, there are no shortcuts, even if the technology seems to offer them. However, some things have undoubtedly been made easier and so, a balanced perspective is maintained.

The book devotes an entire chapter to 'How to Do Things Properly' wherein recording techniques, microphone techniques and mastering techniques are explored in detail. This is an important chapter as it builds upon all that has been revealed and discussed up until this point. The reader therefore has an understanding of *why* such techniques are important. The importance of music to civilisation itself is also discussed. At this point, it becomes obvious why recordings made in the 1950s and early 1960s sometimes sound better, in certain respects, than the equivalents made today. It is an interesting point to ponder amid all of our modern technology. A further point of interest is the education of audio engineers and how this has also changed dramatically since the early post-war days.

In conclusion, this book is unlike any other, being both a practical guide and a very interesting read in its own right, as it covers the history of recorded sound and the key developments which have changed the direction of how we capture audio, both in the studio and at live events. Every aspiring or even experienced audio engineer and producer should have a copy to hand. It may help them to think again about the practice of recording high-quality audio and, consequently, capturing musical performance accurately. This is why we called those black discs with a hole in the centre 'records', a record of a unique musical event which only ever occurred once. Capturing audio accurately is therefore important. The practice of doing so involves a mixture of art and science. This is why this book is, in itself, important. It allows for the understanding and development of both as the reader navigates through the various chapters, understanding what have been useful developments and what have proved not to be.

This book is dedicated to the memory of all the great audio engineers, producers and conductors who, in the period between the late 1940s and early 1960s, produced some outstanding recordings which, in many ways, have never been bettered. They were feeling their way with what were then new tools and developed techniques which remain perfectly valid today. They understood what orchestras and other musical ensembles really sound like in situ and did their best to capture that sound accurately. In particular, the work of Alan Blumlein is acknowledged and celebrated for its impact upon recorded sound (among other things). The value of the legacy left to us from these early years is incalculable. *Audio Technology, Music, and Media* celebrates this reality and brings us forward to the digital age, explaining how we might, similarly, contribute to this legacy. We have come a long way, but we must ensure that our future path is straight and true. This book may act as a guide for those taking us into the future.

Julian Ashbourn

Contents

Chapter 1
How the War Changed Audio

Wars have a convention of accelerating technology, military technology, of course, but also related and supporting technologies. Very often, these improvements in technology quickly find applications within the private sector. In the case of the Second World War, such changes affected, among other things, the way in which we recorded and replayed sound. The technology that was available in this context after the war set the pattern for the next 50 or so years, affecting both the music and audio industries accordingly.

Recording audio, whether music or dialogue, before the war involved the performers playing or talking into a gramophone like horn. This horn captured the sound waves being produced which, in turn, vibrated a diaphragm, which was attached to a stylus which cut grooves directly into a blank shellac/wax-type disc. This relatively soft disc would then be electroplated, creating a more robust master from which other discs could then be reproduced and distributed. However, relatively few people had gramophones to start with, and so the market was not that large and to replay previously recorded material was, itself, a novelty. The audio quality of records made in this way was not terribly good, with an exaggerated middle frequencies and a lack of low and high frequencies. Nevertheless, owning a gramophone and a few discs made in this way was something to aspire to. A sunny afternoon picnic, driven to in an open topped car and laid out in a scenic location, complete with wind-up gramophone, would have been a lovely thing in the late 20s or early 30s. Orchestral pieces would typically require more than one disc, and so periodic changes would have been the norm, but also a part of the charm of owning a gramophone.

There were variations on this theme of direct to disc recording, some of which were especially developed for use in the field. An Englishman named Cecil Watts had developed a direct to disc methodology using a different type of disc material which enabled it to be replayed immediately without going through the intermediate process of electroplating and creating a master. This worked well enough, although the discs tended to wear out rather more quickly. One advantage though was that all

J. Ashbourn, *Audio Technology, Music, and Media*,
https://doi.org/10.1007/978-3-030-62429-3_1

the equipment needed to cut the discs was small enough to be transportable, for example by car, and could thus easily be carried to a particular venue or occasion. In England, the BBC quickly saw the potential of this idea and worked to produce their own truly portable device which could be used in the field. This device, driven by a clockwork motor, could be taken into the field by war correspondents and recordings made right in the battle zones and then sent back for potential broadcast by the BBC news service. It was a beautifully made piece of equipment housed in a wooden box with a hinged lid, in which could be stored blank discs. A very basic microphone could be attached by a lead and pointed towards the desired sound source. One might posit that this was the beginning of what would eventually become electronic news gathering (ENG), an industry in itself. It also occurred to broadcasters that the idea of recording at source and then broadcasting at a later date was an attractive one.

There were several variations upon disc cutting and replay, but there were other ideas in circulation for recording sound. The concept of recording onto a linear medium had been around since 1899, when Valdemar Poulsen introduced the 'Telegraphone' a rather crude device with a very basic, carbon microphone which recorded onto wire and could only produce very low outputs. It was interesting, but not very practical. However, in the late 1920s, another device appeared which used steel tape, running at 5 ft. per second, and whose large reels could accommodate up to 20 min of recording. This was the unlikely named Blatterphone, which resembled a shortened Victorian lamp post with two reels balanced precariously near the top. Interestingly, this seemed to work tolerably well and was widely adopted, if only for reasons of curiosity, by various broadcasters in the 1930s. The problem with the Blatterphone was, although it worked, the sound quality was quite poor with a narrow bandwidth and the tapes had a tendency of breaking, sending razor sharp coils of steel tape, billowing out across the floor. The tape heads were also fragile. In 1935, a variation of this design was produced by Marconi-Stille, which was very much better, although remained far from perfect.

A seed had been sown however for the idea of recording onto tape. This was very attractive to broadcasters who could record a programme for later transmission. Bear in mind that, in those days, practically all broadcasting was undertaken 'live' with no opportunity to correct errors. This possibility did not escape Adolf Hitler, as he appreciated that his dynamic speeches could be recorded and then broadcast many times on different frequencies, to different audiences. The idea of further developing tape recorders was consequently encouraged in Germany, throughout the 1930s and the subsequent war years.

At the Berlin Radio Show in 1935, AEG demonstrated the first really practical tape recorder, which they designated the K1. This was actually a huge leap forwards for two reasons; firstly, its ergonomic layout would be recognised today by anyone who has used an open reel tape recorder, with the tape threaded through rollers and pinch wheels in order to be drawn across a stationery head block. Secondly, the tape itself was manufactured as a layer of ferric oxide attached to a polyester base, not unlike the tapes that would be used for many years in the field of recording. The K1 was quickly followed by a revised machine which introduced, among other refinements, the concept of an AC bias signal being applied at the time of recording. This technique increased high-frequency response and made recorded sound almost indistinguishable from live broadcasts. The Nazis used these machines regularly to broadcast propaganda speeches, and many German radio stations were equipped with them. That the AEG machines were well constructed, with particularly robust motors, was not surprising as the company started out making transformers and electric motors.

When the war ended, the allies found several of these AEG machines, and some were brought back to America and Britain. They were examined by various companies and institutions. In America, Ampex had obtained one and was interested to develop their own version, but this needed funding. Bing Crosby, always with a business ear to the ground, heard about this and privately invested a considerable amount of money in Ampex as he was very keen on the idea of recording something once and having it played back many times, in many locations. Other musicians would quickly see the benefits of this, and, in America, there was a drive towards this end. Consequently, the first practical Ampex machine was installed in many broadcast studios. This established Ampex as a supplier of studio-quality tape machines, a reputation that ran right to the digital age.

In Britain, the AEG machine was picked about by EMI, who saw the potential, but thought that they could build something much better, and EMI still had the right people to do it. Thus, the magnificent BTR1, BTR2 and BTR3 machines were designed and built in quick succession. If any piece of electronics could be said to have been 'built like a tank', it was the BTR machines. BTR, by the way, simply stood for British Tape Recorder. BTR1 was the size of a very large washing machine, with the tape spools and head block on the top plate and two front opening doors which, upon opening, revealed a variety of configurations. Firstly, there would have been just a valve (tube) preamplifier and various connections, but EMI were very forward thinking, and this changed according to model and the client's wishes. Another version displayed two racks, one on top of the other, each containing slots for card modules, each of which would feature a particular set of configuration options. Again, this was very advanced thinking from EMI, who were defining the future for professional tape machines. Yet another option to be found behind the two front doors was a patch bay, enabling custom configurations to be made by the operator by plugging leads into quarter inch sockets. Some leads were fixed at one end, others truly floating, in order to configure the electronics however the operator wished. All the way along, EMI were thinking about the practical side of recording.

BTR1 was a full track machine (the tape had one track across its full width to match a monophonic tape head) which sounded wonderful. From the operators' perspective, there were one or two idiosyncrasies that they found frustrating. The tape path was a little awkward and the head block faced away from the operator, making editing a little more difficult. These points were quickly addressed with BTR2 which was installed in many broadcast and recording studios. BTR3 continued the innovations by offering either two or four tracks. Naturally, the two-track version was perfect for recording stereo, and many fine recordings were made on these machines throughout the 1950s and even in the 1960s. In short, EMI's BTR tape recorders were excellent examples of good engineering coupled to innovative thinking.

It is worth noting the role of the early 'tape men'. They operated separately from the audio (and film) engineers and producers, with the tape machines often housed in separate rooms. The tape men looked after every aspect of recording and playback, configuring and maintaining the machines themselves, managing the storage of tapes and, very importantly, editing. Editing was a physical process of cutting and splicing the tape on a metal block, usually attached to the recorder. They would align the tape over the playback head at a precise point, mark it with a special chinagraph pencil and then perform their highly skilled editing operation accordingly, joining the audio tape back together with clear sticky tape. Skilled operators could remove or replace single syllables, and they were often required to do so for recorded interviews. It was a different but very exciting world in those early days.

At the beginning of the war, there were very few practical tape recorders. Those that did exist were clumsy and unreliable and required very careful handling. After the war, we had a short period of rethinking, based mostly upon the captured AEG machines, and then the birth of the modern, practical tape recorder for professional use. In America, Ampex (and a few others) quickly got into production and started to appear in broadcast and recording studios across America, being particularly welcomed by musicians and singers of the time. In England, the wonderful BTR machines appeared in their standard EMI green paint work. They were often called 'The Big Green Machines' or even 'The Big Green Monsters', but they sounded fabulous and were a joy to work with. Hundreds of notable recordings were made on these machines, in both mono and stereo, and the BTRs continued in use, even when more technically advanced machines were available. They were truly a milestone in the history of recorded sound.

Chapter 2
The 'V' Record Label for the US Troops

Entertainment was important throughout the war, both in the affected countries and especially for troops based overseas. Away from home and, sometimes, having to fight gruelling battles, entertainment was an essential morale booster for these men.

The British, in true idiosyncratic British manner, conceived the idea of concert parties. Small groups of men would put on shows and perform music, travelling around from one battle zone to the next, never tiring of their task. Famous singing stars such as the wonderful Vera Lynn would often go with them, taking the same risks as the soldiers whom they entertained. Those troops based in England were further entertained by a young lady from Lancashire, who had previously been a GPO telephone operator, but who sang like an angel. She travelled the length and breadth of Britain, sleeping in trains, in rough digs and grabbing a meal anywhere she could. Her name was Kathleen Ferrier, and she worked tirelessly to entertain troops stationed in Britain. After the war, she had a relatively short but glittering career, singing a wide repertoire, from traditional English folk songs, to works from Mahler, Brahms, Schubert and others, before being cut down by cancer at the age of just 44 years. Kathleen's naturally warm and sincere personality, coupled to a lively sense of humour, ensured that she was much loved wherever she went. Her beautiful voice came from within a beautiful person.

The Americans, who had of course entered the war much, much later, did things differently. They simply took record players with them and used them wherever they were stationed. This meant that they could listen to the latest songs from back home and, consequently, not feel so isolated. There were complications at first because American musicians had gone on strike, refusing to record for any of the established record labels as they were not receiving royalties on sales, an understandable objection. However, they did agree to record music for troops based overseas, providing that the resulting records were never made commercially available. And so, a new record label came into being. This agreement was largely brokered by a recording enthusiast, now in the army, Lieutenant George Vincent, who was

© The Author(s), under exclusive license to Springer Nature Switzerland AG 2021
J. Ashbourn, *Audio Technology, Music, and Media*,
https://doi.org/10.1007/978-3-030-62429-3_2

able to convince the four major record labels of the time to allow their artists to record for military purposes.

The V-Disc was initially the brainchild of Captain Howard Bronson, who realised that a morale boosting initiative was needed for military personnel serving overseas. He had been assigned to the Army's Recreation and Welfare Section and was thus in an ideal position to make the V-Disc a reality. At first, the discs were mainly of military 'motivational' music and recordings from shows and films. This in itself was popular enough, but the men missed home and listening to the popular bands of the day, so the agreement that George Vincent pushed through was an essential element in the ongoing success of the V-Disc. In addition, radio networks sent feeds and live broadcasts to the V-Disc headquarters in New York, allowing excerpts from live shows to be recorded as well. Ironically perhaps, the discs themselves were pressed by major record labels such as Columbia and RCA Victor, and after the war, these stampers and masters were destroyed, together with the existing stocks of V-Discs. Servicemen were forbidden to take the V-Discs home although, inevitably, many of them would do so but, if caught, they would face heavy fines.

The discs themselves were 12 in. in diameter and ran at 78 rpm. This allowed for around six and a half minutes of programme, whether spoken word or music. Originally, they were made of shellac, as all 78 s had been prior to the war. However, this material was brittle and, in any case, became hard to source during the war. Consequently, a special mix of vinyl (actually a mixture of Vinylite and Formvar) which was much more durable and could easily be shipped out in quantity. The organisation involved in creating and running the distribution network was impressive, as was the selection of artists and repertoire and, of course the recording sessions themselves. It created a pattern that would be used to very good effect after the war had ended.

And so, the V-Disc was extremely successful, but in more ways than originally foreseen. It established a large-scale manufacturing and distribution network, proving that this was entirely viable. Importantly, it also established a thirst for recorded music, proving that playback was straightforward and reliable enough, even in war zones. The American servicemen had become used to receiving records of the latest music, whether it be orchestral, big band or jazz, and the playing of these records had become a part of their relaxation and social activities. No doubt, allied soldiers and airmen had also been exposed to the V-Disc, in one way or another. So the playback of recorded music had grown from a plaything of the few to an expectation of the many, and after the war, a readymade market therefore existed.

Socially, the V-Disc was also important as it brought people together under a common interest. Music is, after all, a great leveller. Furthermore, it exposed many to different types of music with which they had not previously been familiar. Jazz was extremely popular, and during the decade and a half that followed the end of the war, many jazz bands were established and thrived accordingly. Some of the musicians involved had recorded on the V-Disc, this being their own introduction to the world of recording and the possibilities that this introduced. This, in turn, enabled a number of independent recording studios to become established. The wide distribution of music during wartime had sown the seed for what would become a massive industry in peacetime. Different record labels flourished, each with its particular roster of artists and, to some extent, each with its particular sound. Indeed, they competed with each other in this respect as the term 'high fidelity' came into being, and each label looked for some unique feature in order to distinguish itself in this context. The microgroove record had appeared, allowing longer playing times which, in turn, allowed for a wider variety of programming. Popular music settled on 45 rpm playback speed, with 33 and a third being used for classical music. And so, the various record labels on both the sides of the Atlantic, settled down to exploit these developments, each one claiming a better sound than the others. One such label that became very interesting was the Mercury Living Presence label. By 1950, it had evolved as a major influence, having acquired the rights to distribute various European pre-war recordings in the US, as well as developing its own catalogue. It employed the services of various independent audio engineers, including Bob Fine, who suggested recording an orchestra (the Chicago Symphony Orchestra in this case) with a single microphone suspended around 25 ft. over the podium. He recorded a performance of Mussorgsky's Pictures at an Exhibition, which was subsequently well reviewed, with one reviewer observing that it sounded as though you were actually present at the recording. This lead to the 'Living Presence' moniker being added to the label name. They would go on to make many landmark recordings.

In Europe, recording labels were similarly striving to make high fidelity records, even though the equipment available to them was limited in many respects. Nevertheless, there existed some wonderful producers including Londoner Walter Legge, who worked mostly with EMI, and German born Bruno Walter who, having become a French citizen, subsequently went to America, but continued to be associated with some European recordings. Bruno Walter worked closely with Gustav Mahler and consequently came into contact with another who appreciated Mahler's music, Kathleen Ferrier. He once said that the two most important events of his life were meeting and knowing Kathleen Ferrier and meeting and knowing Mahler—in that order.

And so, the V-Disc was an important herald for what was to come. It proved that it was viable to produce and distribute recorded material on disc, upon a wide scale. That, in turn, inspired a thriving recording industry after the war, ensuring that disc recording and playback technology would be developed to a high standard and become something quite different from wind-up gramophones and 78 rpm shellac discs. The servicemen who enjoyed receiving their supplies of V-Discs were, unwittingly, the prototype audience for the forthcoming high fidelity boom. Those that returned would enjoy listening to better quality sound and a wider range of music. The ancillary equipment required for playback would also be developed into something quite different, with amplifiers and loudspeakers all pushing forward in their own technical development and, of course, this post-war boom in audio reproduction would eventually play an important part in the Japanese economic miracle. The V-Disc thus deserves to be remembered fondly (Figs. 2.1 and 2.2).

Fig. 2.1 The V-Disc cover

Fig. 2.2 A typical
V-Disc label

Chapter 3
Stereo Sound, Film Sound and the Legacy of Alan Dower Blumlein

In September 1924, the International Western Electrical company (IWE) took on a new employee, a young man of 21 years who had no practical experience but who had earned himself a first-class honours degree in electrical engineering from the City and Guilds College, which he had entered on a Governor's scholarship. It was a 4-year course, but the young man had completed it in 2 years. His name was Alan Dower Blumlein. He worked initially on long-distance telephone lines and other related projects. In particular, he looked at reducing crosstalk over long lines and successfully accomplished this. By 1929, he had already been awarded seven patents in this area. He was clearly an exceptional individual, who adopted a solid scientific approach to everything he worked on, taking great pains to understand the relevant principles involved before producing innovative solutions to given problems. His response to solving the crosstalk issue was to realise that it was not the cables themselves that were the main problem, but the coupling transformers at either end of the line. And so, he simply designed radically new transformers, which alleviated the problem and which were quickly adopted. This was a part of Blumlein's genius, the ability to focus on an issue in minute detail, until he really understood the principles involved and what was going wrong.

In March 1929, Blumlein joined Columbia Gramophone, which subsequently merged with The Gramophone Company to become EMI, for whom he would produce an astonishing 121 patents. For EMI he worked on many projects, always bringing fresh and innovative thinking to the attendant issues. EMI were themselves quite adventurous and were already starting to look at the possibility of television. HMV had already demonstrated a mechanical system giving 150 lines of resolution, which was considered very high. Given that cathode ray tubes of the time were rather primitive and that there were no electronic cameras, the problem of television represented quite a challenge. The Television Advisory Committee had set a target of 240 lines of resolution, but EMI's director of research at the time, Mr. I Schoenberg, decided to adopt 405 lines as the standard to work towards and the team, including Blumlein, set to work accordingly. The high-definition service that

J. Ashbourn, *Audio Technology, Music, and Media*,
https://doi.org/10.1007/978-3-030-62429-3_3

9

was launched in 1936 continued in use until 1950 and, indeed, beyond. Blumlein addressed the problem of the early CRTs losing focus, with some typically innovative circuitry, ensuring that the resulting design could be mass produced. He seems to have worked on nearly every aspect, and it has been suggested that, upon adopting Alexandra Palace as the new transmitting station, there was hardly anything in Alexandra Palace that had not benefited from Blumlein's expertise, from scanning systems to signal transmission, to stabilising the tubes to be used in television cameras. Blumlein was never afraid to flout convention where necessary and, if existing methodologies did not seem right, or reliable enough, he would simply design anew. It seems that EMI had inherited a gem of a scientist. However, the true brilliance of this gem was yet to be revealed.

With the war years upon them EMI worked on RADAR, and unsurprisingly, Blumlein was an essential member of the team. The precision that RADAR demanded was difficult to achieve with existing techniques, which simply did not have the granularity required. Blumlein looked at one aspect after another, redesigning where necessary and innovating new approaches as he went. While many scientists worked on RADAR (which was also given to America, where work continued), they all seem to agree that, without Blumlein, there would have been no RADAR. He was the catalyst that turned ideas into reality. Having perfected ground to air and air to ground RADAR (the latter being so refined that it could actually be used to create maps), Blumlein and others were working to improve air to air RADAR when a test flight on which Blumlein and two of his EMI colleagues were participating on 7 June 1942, got into trouble and the Halifax bomber that they were travelling in crashed. There were no survivors. That incident was such a loss to the war effort that it was commented on within The House of Commons. It was not just a loss to the war effort however, but a loss to Britain and, indeed, to the world.

Within Alan Blumlein's short career, little more than a decade, there were 132 patents in all granted to him, with or without collaborators. But he was seldom seen in public and was not inclined to write scientific papers. This was primarily because he was always so busy. He did have a private life and enjoyed, among other things, attending music concerts. These experiences no doubt caused him to think about music reproduction both in the home and on film, and how it might usefully be improved. The invention of Stereo (sometimes referred to as Stereophony or Binaural) sound recording and playback was a masterpiece. Blumlein's starting point, as always, was to understand the principles involved: how we heard sound from one or more point sources with both ears, the nature of sound waves at different frequencies and more. He understood the distinction between how we hear low frequencies and how we hear high frequencies; the former is a matter of time difference and the latter one of intensity difference. He also recognised that, if we had multiple loudspeakers in a room, the listener would hear the directional output from all of them with *both* ears. This was fundamentally different from the experience of listening to two or more channels via headphones. And so, he set about designing a system that required just two channels and two loudspeakers.

The solution was based upon the sum and difference output from two microphones. At this point, the only microphones available to Blumlein were omni-directional, not ideal for the purpose, although he still made it work, giving results analogous to two cardioid microphones pointing in opposite directions. He then went on to define moving coil microphones and dramatically improve replay systems, recommending a different material for records (cellulose acetate) and a different type of stylus (sapphire) as well as specifying how stereo records could be cut using the 45° angle for modulation which was subsequently adopted worldwide.

Even though Blumlein had made quantum leaps in the recording and reproduction of music in the home, it was film that he was really thinking of and, eventually, stereo film tracks would become the norm. He even adapted the idea for monitoring the presence of enemy aircraft. Blumlein's patent no. 394,325 on 'Improvements in and relating to sound-transmission, sound-recording and sound-reproducing systems' ran to 22 pages and made 70 separate claims. Most would be happy with one or two specific claims, but Blumlein did things properly and defined the entire stereophonic recording and replay chain. He was at least 20 years ahead of his time (this was in 1931) and was even considering multi-channel recording and playback for specific purposes. Simultaneously, he was working on many other crucial projects for EMI. How lucky they were to have inherited him.

Properly configured, a stereo recording will capture a wonderful, three-dimensional sound field, from just two microphones placed in a coincident or near coincident configuration. All instruments within this sound field will appear in their precise positions, the listener hearing an orchestra or choir, exactly as it was in the original, physical space. Width, height and depth information will all be captured. Furthermore, the complex phase and reverberation characteristics, both between the instruments and between the instruments and the enclosed space in which they were playing, will all be captured intact. This is an important aspect of true stereo recording that many do not understand.

In the early days of stereophonic recording, there were enthusiasts who could reliably ascertain, from the recording alone, the venue in which the performance took place. Those who have practised critical listening will find that they can do the same after a while. This is because the sonic signature of the concert hall or other location is reliably captured and, if the recording engineer does not tamper too much with the sound, these sonic clues remain in place. This is possible with these early recordings as they were not 'mangled' by the many studio practices that take place today.

The stereo enthusiast in the 1950s and 1960s had a wonderful choice of good quality equipment from which to choose, all of which, while not at bargain basement prices, was at least affordable. This is very different today with products that are any good at all, financially well out of reach of the average listener. In Britain, Europe and America, the great names in audio were being established. In Britain alone, we had Leak, Wharfedale, Garrard, Tannoy, Radford, Lowther, Rogers and many others. The author once owned a pair of original Lowther Acousta loudspeakers, with single full-range, paper-coned drive units placed in horn-loaded enclosures. They sounded fabulous, with a full frequency range and lightning fast response to transients. They were also superefficient and could be driven by a few watts of power. I have never heard loudspeakers which were as convincing as these since. As an aside, it was Paul Voigt who designed the original horn-loaded cabinets and Donald Chave was the chief engineer who, after Paul had migrated to Canada in 1950, developed the PM1 and later drive units. These had massive magnets and light, paper cones which provided the fast transient response. The horn-loaded enclosures ensured that a full range of sound was projected into the room. The early enthusiast had good quality valve (tube) amplifiers from which to choose and a wealth of loudspeakers. Early set-ups typically included a reel to reel tape recorder of some kind or another. Manufacturers such as Ampex, Revox and the wonderful Ferrograph company provided the goods, and they were all of high performance and sound quality. I know of Ferrograph machines that are still being used today; they were extremely well made.

All of the above was made possible by Alan Blumlein's invention of stereo sound. It was the key that unlocked a global industry, initially focused mainly in Europe and America and later in Japan and the Far East. It also brought a revolution in film sound, enabling cinemas to create realistic effects and place actors both visually and acoustically. This, in turn, lead to a very complex film production industry which is still thriving today. Although, it took some time for the film industry to catch up with proper stereo sound.

The transition from silent films to 'talkies' was not an easy one. At first, sound was added to makeshift movie theatres simply by playing gramophone records behind, or adjacent to, the screen. The synchronisation between sound and picture relied upon the dexterity of both the projectionist and the sound man. Naturally, the result was hugely variable. In the 1920s, companies such as Western Electric were using discs but experimenting with recording the sound track straight to film using, mostly, a variable density modulation method which was also variable in quality, depending upon how expertly the film was processed. If the resultant film was either underexposed or overexposed, this would also affect the sound track, making it weaker or stronger accordingly. At this time, there seemed to be dozens of people experimenting with sound for films. Many systems were so cumbersome that they quickly became unrealistic in use, especially if they required any particular synchronisation between the recorded sound and the recorded film.

The answer was obviously to place the sound track on the film, but how? In 1922 Dr. Lee de Forest developed a special photoelectric cell for recording Movietone optical sound tracks on film. The UK rights to the de Forest 'photofilm' were bought by Mr. C.F. Elwell, who continued development of the system in his premises at Cranmer Court in Clapham. The British Acoustic sound system worked better by placing an area-modulated audio track on a separate 35 mm film and synchronising the two together for playback. This was in regular use into the late 1930s. Actually, the obvious answer, recording a stereo, area-modulated track on the same film, had already been achieved in 1935 by Alan Blumlein for demonstration purposes, but no one seemed to realise the potential. It was not until the late 1950s that this became a reality. This is partly because the attention of the film industry was diverted by the quest for colour, for which numerous experiments were being made.

The next significant step was when Dolby Laboratories began to look at film in the 1970s and realised that, in addition to a stereo, area-modulated track, you could write information to the blocks of film between the perforations. This enabled noise reduction and, eventually, special audio effects to be written directly on the film and of course, as it was all on the one film, there were no synchronisation issues. However, cinemas had to be equipped with the right equipment in order to make use of these developments, and this took some time. But then came the digital revolution.

Digital cameras brought film making to the masses of amateurs who wanted to get into the film industry. Film was still being used, but for news gathering, TV and other purposes video and then digital recording to solid state media became very popular. Amateurs can now make very good quality short digital films. But sound remains an issue. This is because even 'prosumer' digital cameras have poor quality in-built microphones which are fixed in position. The answer is to record the audio separately (as still happens for major feature films) and synchronise the two together after the event using specialised software. This is made possible by small digital audio recorders which may be mounted beneath the camera on a solid tripod and connected directly to the camera's sound system via a line input socket. The aim here is not to record high quality audio onto the camera's sound track but, at the beginning of a recorded section, place what is known as a 'slate' tone onto the cameras sound track, together with a low quality audio track. The high quality audio is recorded onto the digital recorder via a series of microphones, usually up to four or sometimes eight. This allows the sound man to record a good stereo track, plus one or more spot microphones for dialogue, often suspended on hand-held boom poles which may be moved around, following the actors. The two digital files, for video and audio, are then synchronised in a software simply by aligning the two slate tones. It is not very precise, but is certainly good enough for making amateur films and is a widely used technique.

Better precision is achieved by using something called 'timecode'. With timecode, the audio recorder and camera are connected together (both must have timecode inputs) by a cable, and one or the other generates a high precision code which is recorded onto both devices. When subsequently edited in software (which must also recognise timecode) the audio and video may be very precisely aligned, down to fractions of a second. Consequently, if whole shots are removed, or special effects added, the two may always be brought back to perfect synchronisation using timecode. Professional films are made this way, and film music composers may also synchronise their music to copies of the film and send them back and forth for audition and acceptance.

Although feature films may have four or more tracks for surround sound, in order to realise playback, one needs the appropriate equipment, and, unfortunately, there are competing standards which make this something of a nightmare. Fortunately, most of the world's greatest films were made with two track audio which may easily be played back on any television. However, there is a distinction between two-track audio and proper stereo sound as defined by Blumlein. Very few films were made in proper stereo, despite how they were advertised, and the same problem continues today. It is a shame, because true stereo provides for a realistic, three-dimensional sound field. Ironically, amateurs may easily make true stereo films by attaching one of the specialist, high-quality microphones already aligned for either coincident or near coincident stereo, to the top of the camera and ensuring that the lens is set to a static zoom position, usually at 50 mm or a slightly wider angle setting. This procedure, with the microphone connected to a linked digital recorder, can work extraordinarily well. In any event, we owe Alan Dower Blumlein a great deal (Fig. 3.1).

Fig. 3.1 Alan Dower Blumlein

Chapter 4
The Physics of Sound

In this chapter, rather than presenting a lot of equations and complex mathematics, the physical properties of sound shall be examined in a practical manner. This will be of particular value to amateur audio engineers and producers (also to many professionals).

Sound and light travel in waves. An analogy often given for sound is that of throwing a pebble onto the surface of a still pond. Waves radiate outwards from the point of impact, just as sound waves radiate from the sound source. This is due to a disturbance in the air around us (if you did this underwater, you would still get radiated energy as waves through the water). If you bang two sticks together, you will get a sound. As the sticks approach each other, the air immediately in front of them is compressed and energy builds up. When the point of impact occurs, this energy is released as sound waves. If you try the same experiment with two heavy stones, exactly the same thing occurs, but you get a different sound due to the density and surface of the stones, and as they have likely displaced more air, a louder sound. And so, a physical disturbance in the atmosphere around us will produce a sound. This is an important point to understand. When two heavy rain clouds collide and rupture, we get thunder, as this is analogous to our experiments, but on a grand scale. The energy released is much greater.

Musical instruments all operate on this principle. Stringed instruments, such as violins, violas, cellos and acoustic basses, produce sound by vibrating a string over a sound box, disturbing the body of air within and thus producing sound waves. If you vibrated the same string without the sound box, it would still make a sound, as it would still be disturbing the air around it, but the waves created would be slight and the sound very low in amplitude (volume or power). The sound box of the stringed instrument has its own resonant frequency and will produce the loudest sound at this particular frequency (pitch) but will still produce sound on either side of this resonant peak. The strings of the stringed instrument also rest on a bridge at the end which is attached to the body (the other end being the fretboard and tuning pegs). The bridge is directly connected to the body of the instrument and usually has

© The Author(s), under exclusive license to Springer Nature Switzerland AG 2021
J. Ashbourn, *Audio Technology, Music, and Media*,
https://doi.org/10.1007/978-3-030-62429-3_4

a post beneath it which connects the back and front of the soundbox together. This ensures that the vibrations from the string are transmitted fully to the body of the instrument. The strings may be bowed, which causes them to vibrate finely, or plucked, which causes them to vibrate in a much cruder manner, producing a quite different sound of shorter duration.

Wind instruments work slightly differently but still rely on a body of air being disturbed in order to produce a sound wave. A clarinet, for example, contains air within its body. When the clarinet player blows into it, this body of air is compressed and finally breaks out through the throat of the instrument as sound waves. The frequency of these waves may be controlled by opening or closing valves on the body of the instrument, which lets more or less air escape from the mouth of the instrument. A saxophone works in a similar manner but has a different sound characteristic due to the construction of the instrument. Brass instruments without any valves at all rely upon the player to vary the pitch of the instrument by his own playing technique, but the principle remains the same. A body of air within the instrument is compressed or modulated until it is forced out of the instrument as sound waves.

A piano produces sound by striking strings, which are highly tensioned, with hammers, operated from the keyboard. The strings vibrate within the body of the instrument and are, in a grand piano, further reflected by the sound board. A concert grand piano sounds a little different from the upright piano in your local saloon, due to a different size of frame and a different body construction, but the principle is the same. An organ however works more like the wind instruments, blowing air through a series of tuned pipes in order to produce a sound. The great organs in some of our churches and cathedrals have literally hundreds of pipes, often grouped together in banks, and can produce a very wide range of sounds. They also have a wide frequency capability, some of them able to produce notes as low as 30 Hz (cycles per second) which is very low indeed. The frequency of a given note is measured by the distance between the peak of one wave and the next. The closer the wave peaks are, the higher the frequency being produced. Humans, in theory, may be able to hear notes within the range of 20 Hz to 20 kHz. In practice, an adult male will be lucky if he can hear up to 15 kHz. But this is still very high. If you look at a spectrogram of an orchestra playing, most of the energy is within about 200 Hz to 9 kHz, with occasional leaps up to 15–18 kHz for some solo instruments. These instruments produce natural harmonics, which can reach beyond 20 kHz, but we shall consider that point elsewhere.

In conclusion, a disturbance of the atmosphere around us which produces an impact of some kind will make a sound, which will be radiated out from the source as a series of waves, diminishing until the disturbance has ceased. These waves may however be reinforced by the topology of the land or structures around them. All musical instruments work on the principle of causing a disturbance and, if possible, amplifying the resultant sound waves by a resonant box or pipe. A human voice works in the same manner. We create a disturbance by vibrating our vocal chords, using air from our lungs, and amplifying the sound via the topology of our mouth. Trained singers are simply using their lungs, vocal chords and mouth as a musical instrument. And a very effective instrument it is.

The pebble in the pond radiates waves equally in all directions until they are interrupted by boundaries of some kind. The sound from musical instruments also radiates in all directions (360° around them), but not in an equal manner. There are various reasons for this. Imagine we have a solo trombone player in a room about 40 ft. by 20 ft. by 10 ft. high, with a musician standing off centre around 15 ft. from the rear wall. Let us choose the wonderful Jack Teagarden, otherwise known as Big T. Let us stand 10 ft. directly in front and let Big T play a 1-min solo on his trombone. Now let us move 90° around in an arc, remaining 10 ft. away, but this time standing by Jack's side. He plays exactly the same piece, but the sound is different. Now let us move to 180° so that we are standing behind the player. Big T plays again, but again the sound is different. Back to the other side, the sound is different again and, lastly, back to where we started from and the sound is similar to the first run through. The different sounds that we encountered are the result of some complex physical properties. When we were standing right in front of Big T, we heard a considerable amount of direct sound from the bell of his trombone. The sound waves were exiting the trombone and making direct contact with our ears. However, we were also hearing a certain amount of reflected sound as the waves bounced off the room boundaries and entered our ears from a different direction. These sound waves were very slightly delayed, and this effect is known as reverberation. So, the sound that we actually heard was a mixture of direct and reflected sound which, together, give the trombone its distinctive voice, especially if it is in the hands of someone like Jackson Teagarden, Big T.

As we moved round to 90°, we heard less of the direct sound and more of the reflected sound. Consequently, the sound was a little less distinct with slightly softer transients and a slight loss of the highest frequencies. At 180°, standing behind Big T, we were hearing mostly reflected sound, although we still knew that it was a trombone being played. At 270°, a mixture of direct and reflected sound, but a different sound to that at 90°, due to the closer proximity of the side wall at that position. So, we can see how our position relative to the player alters the sound that we hear, and this will vary according to the instrument being played, the size of the player and so on.

Another characteristic that presents itself here is phase. As we moved around Big T, the direct sound that we heard reached our ears at a very slightly different time. When we were standing behind him, the direct sound waves, which radiate out from the bell of the trombone in all directions, had to pass around and through the body of Mr. Teagarden. Consequently, they were slightly delayed. If we took a waveform of the direct sound heard at each quadrant and overlayed them, we would see that they shifted slightly in position. This is the effect of phase. This is why, if you reverse the connections to one of your loudspeakers in your hi-fi set-up, the sound becomes indistinct. That is because the cones in one loudspeaker are moving forwards while the cones in the other are moving backwards. They are out of step with each other or out of *phase*. This is an important factor when it comes to recording audio, as we shall see later.

Now let us introduce another trombone player into the room. This time we shall choose Bob Brookmeyer, who liked to play the valve trombone. We shall place him in line with Big T and off centre in the other direction. Now we shall ask them both to play the same piece of music and to stand in the centre of the room about 10 ft. in front of the players. The sound we hear is very different. Even if the two players had identical timing and phrasing, the sound will remain very different. Much richer and certain notes will seem to become more emphasised. This is because there is much more energy being pumped into the room, with approximately double the pressure of sound waves and, of course, twice the amount of reflected sound bouncing around the walls of the room. There will also be a slight phase difference in the sound that we hear between the two players, as we are only roughly in the middle and one of them will no doubt be playing more loudly than the other. We may repeat our earlier experiment, with interesting effects as we can no longer be always at the same distance from the two players.

Now let us introduce a third trombone player; this time we shall choose Kai Winding who had a different style and tone, and we shall place him between the other two players, but 5 ft. behind them. Let them all play the same piece again, and we shall hear yet another, quite distinctive sound. The sound we hear is very rich and very complex. This is because we are hearing a mixture of three fairly direct trombone sounds, plus a very complex pattern of reverberation. The phase and reverberation characteristics, between the musicians and between the musicians and the space in which they are playing added to the direct sound of their instruments, are what an orchestra, chamber group, choir or other ensemble give, their particular sound.

One reason why Alan Blumlein's invention of stereo was so important is that if you record true stereo, as defined by Blumlein, you capture all of this spatial information in the recording. In particular, all of the phase and reverberation characteristics remain intact, and the special sound that one hears at the podium position is preserved. The orchestra sounds like an orchestra. Unfortunately, the majority of modern recordings do not do this. They, instead, record with multiple microphones and mix the signals via a mixer, artificially using pan controls to place the instruments roughly in the position that they should be. With some types of music, the spread between the two channels is completely artificial with instruments placed anywhere that the audio engineer decides.

To conclude, sound waves radiate outwards from the source and, if not interrupted, will spread equally in all directions until they decay naturally. This starts to happen as soon as the sound has stopped. If a boulder should become dislodged and roll down the slope of a hill before finally hitting another, stationary rock, there will be a loud crack, with sound waves radiating out through 360° and also upward into the air. The ground will prevent them from penetrating too far downward, although they will go some way, and the hill will prevent them from going too far in that direction as it will absorb some of the sound. In all other directions, as soon as the sound of impact has ceased, in this case quite quickly, the sound waves will stop being generated at source, but those already generated will slowly decay as they lose their energy. Just like the pebble being thrown into the pond. Sound waves can however pass through obstacles such as building walls and other constructions. When a sound wave meets an obstruction, either it will pass through it and exit in a diminished form or the obstruction will totally absorb the sound. Concert halls and recording studios are designed, using a variety of materials, in order to control the behaviour of the reflected sound waves, and thus ensure a good primary sound quality.

Indeed, the knowledge of how sound behaves is required for the design of nice sounding halls, but also for audio components such as microphones, which may exploit these known characteristics to good effect. This is why microphones have different polar pattern characteristics and are designed to be more or less directional. The design of loudspeakers also benefits from the knowledge of how sound behaves within an enclosed box and the different fundamental types such as infinite baffle, bass reflex, transmission line and horn loaded, all exploit this knowledge. Unfortunately, loudspeaker designers do not know what kind of room their design will end up in, which is why there is no guarantee of a good sound in situ for any loudspeaker, one just needs to try and see.

The other parameter of sound waves is their amplitude or energy. The larger the amplitude, the louder the sound as the distance between the peak of the upper half of the wave and the trough separating the waves increases. As the wave slowly decays, the amplitude decreases. This is the dynamic range of the sound in question. For an orchestra, the dynamic range is the difference between the quietest sound and a full crescendo. The human ear can accommodate quite a wide dynamic range, although we should be careful and never expose ourselves to abnormally loud sounds. Recording a full dynamic range accurately can be a little complex, and as audio recording technology progressed, the ability to capture dynamic range in an uncompressed manner increased. Today, we may easily capture the full dynamic range of an orchestra (Fig. 4.1).

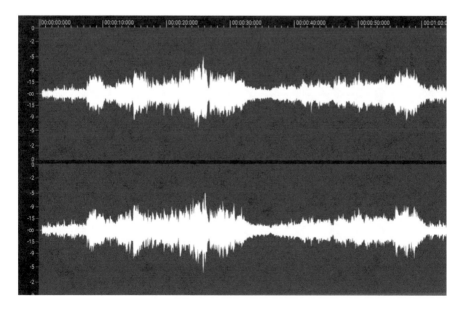

Fig. 4.1 A spectral wave view of a burst of music

Chapter 5
The Advent of Tape and Moving Coil Microphones

The development of the tape recorder, during and especially after the war, was to have far-reaching consequences for audio in general. At first, it was of interest mainly to broadcasters who saw the potential for recording a programme and being able to edit it before it went out over the air. This alone was something of a revolution in broadcasting, and the practice was quickly adopted on both the sides of the Atlantic. In addition, the ability to play back the programme material multiple times without the original artists being present was appreciated.

The record industry soon pricked its ears up and realised that recording to tape, editing where appropriate and then using the master tape to create the record stamper gave them many advantages. At first, in both camps, all this was happening in glorious monophonic sound, and it took quite a while before stereophonic sound really took off. Undoubtedly, one reason for this was the availability of good quality stereophonic tape recorders.

In Britain, Blumlein's work on stereo had been understood since 1931, but the available hardware was not really ideal. It was Blumlein's specifications for moving coil microphones that really made the difference. Up until then, microphones had been rather crude carbon types with a limited frequency range. With moving coil microphones however, the diaphragm could be made small and light, giving considerably better high-frequency response. Furthermore, the polar pattern, that is, the directional acceptance of sound at the diaphragm, could be controlled. Microphones could be made to be omni-directional or unidirectional, giving them the potential to be pointed at the sound source while rejecting sound from the sides and from behind. Eventually, the popular cardioid polar pattern (named because it was heart shaped) was widely adopted and remains so today. But there were also figure of eight response shapes which accepted sound from both in front of and behind the microphone. If two figure of eight microphones are configured as a coincident pair (this means that the diaphragms are more or less vertically aligned), then a variation of Blumlein's stereo picks up sound fields from both in front of and behind the microphones. This configuration is still known as a 'Blumlein pair' today. If used to

J. Ashbourn, *Audio Technology, Music, and Media*,
https://doi.org/10.1007/978-3-030-62429-3_5

record a classical music concert, it captures both the orchestra and the audience, giving a very realistic impression of actually being there in the concert hall.

But we are jumping ahead a little. First came the widespread adoption of two channel sound. I refrain from using the word 'stereo' because many audio engineers had not grasped the importance of true stereo and were simply putting up two microphones somewhere in front of the sound source. Actually, a variety of configurations were being used. In America, a popular theme was to use three microphones arranged in a line in front of an orchestra with a simple-level mixer alone adding a proportion of the centre signal to the two side signals. It worked well enough. Some experimented with more microphones, and in any event, the results were interesting enough for record labels to claim that their particular way of recording was unique, and there were many variations, with interesting names, mostly claiming stereophonic sound, even though they were not recording true stereo.

Sometime around the year 2000, the author initiated a campaign on behalf of the online audio resource TNT Audio, named 'Save Our Stereo' (SOS). As I recall, there were well over a hundred organisations and audio engineers who signed up in support, displaying the SOS logo on their own websites. I do not know what has happened to them all, but there were still people around then who understood the distinction between stereophonic and multi-channel recording. Today, I think just a handful of specialist labels and independent audio engineers work this way. This is a great shame, as many will never actually hear stereo sound, no matter what is written on the cover.

The tape recorder went from strength to strength. Domestic machines were robust and sounded very good. Professional machines more so and then, a jump to four track recording took place. This enabled recording artists to 'over dub' on their own recordings, enabling a variety of effects. This was quickly exploited within the popular music field, and many musicians bought these machines for their own project studios. In professional circles, tape width increased from a quarter of an inch to half inch, and eight-track machines started to appear. Then there was 1-in. tape and 16-track recorders until the market seemed to stabilise for a while on 2-in. tape and 24 tracks. Thirty-two-track recorders even appeared towards the end of this period. It seemed that the whole industry was obsessed with having large numbers of available tape tracks. This was both a blessing and a curse, as far as audio was concerned.

Of course, multi-track tape recorders required multi-track mixers in order to record and later mix down the tracks to a two-track master tape. These mixers were large, complex and very expensive. Indeed, equipping a new sound studio in the 1980s or early 1990s would be a very expensive undertaking. But some of these mixers were masterpieces of analogue audio design. In Britain, companies such as Neve and SSL made superb mixers that sounded fabulous and found popularity in studios around the world. In America, mixers from companies such as Harrison made large mixing consoles that found favour in television and film studios in particular, due to their individual sound. These mixers became huge because, in addition to their 24, 32 or more input channels, they featured a number of group and sub-group channels, typically up to 16, plus, of course, the two master channels. The group channels enabled combinations of the main input channels to be collected together in groups, in order that they may more easily be managed. After all, no audio engineer can realistically manage 24 input channels in real time. A fewer number of group and, finally, sub-group channels made things much easier. They might also use group controls to send a signal out to headphones worn by the musicians. Each input channel would typically feature a comprehensive equalisation (tone control) section, usually with parametric bandwidth control and additional shelving high and low filters. The equalisation section on Neve consoles was particularly favoured by some. There would also be channel inserts, where outboard effects such as compressors, limiters, delay lines and so on could be inserted directly into the channel. And there would be a group of 'send' controls to send the signal within the channel to outboard processors such as reverberation devices. Each channel could switch between microphone or line inputs and would have switchable 48 V phantom power for microphones. A 'trim' control at the top would match the channel to the input device, and at the bottom, there would be a fader for level control and a potentiometer for pan control (to position the signal between the two channels). Other facilities might be provided upon request. The mixer would also have an array of inputs and outputs. No wonder they were heavy and complex.

The problem with all this circuitry is that as the input signal passes through each section of electronic circuitry, it is slightly degraded. This usually manifests itself as a softening of transients, a slight loss of high frequencies and occasional phase issues. Indeed, a mixer which could remain linear in phase from input to output would be unusual. Remarkably, some of these monolithic electronic devices would sound very good and some musicians, particularly in the popular music field, would have preferences in this respect and only use studios which had a particular mixer. This was not really logical as each of these monsters were handmade, and so each one, even of the same model type, would sound slightly different. This was also noted by some who would insist that individual mixers sounded different.

However, there are many other variables in the chain, starting with the microphone. In some respects, microphone design has not changed that much over the years and some manufacturers, such as the German company Neumann, make a virtue of this, especially with their popular U47 large diaphragm design which has remained popular with vocalists for decades. The fact is that the fundamental principles of microphone design have been understood since the 1950s. Today, some of the circuitry has been improved, and even mid-priced designs can sound very good, that is, add very little sound of their own. But all microphones add *some* sound of their own to the signal. Ironically, some of those that add a distinct sound of their own are favoured for this very reason.

Most microphones today are described as either dynamic or condenser types. Dynamic microphones work by pressure on the diaphragm moving a coil within a magnetic field, creating a voltage analogous to the sound being picked up. Microphones of this type are popular with touring musicians playing at live venues, due to their ruggedness. Condenser microphones are generally of a higher performance but require power, due to their design whereby the diaphragm forms one half of a capacitor, the other being a static plate with a voltage across it. Some affordable condenser microphones are powered by an internal AA battery. Professional microphones are mostly powered by 48 V pulled up from the mixer via the microphone cable (phantom power). This is possible because the cable is of a three-pole 'balanced' configuration, and so the voltage does not interfere with the signal. The design of the microphone diaphragm, including size and material, will determine how the microphone responds to transients, its frequency response and its power bandwidth. A small size is usually faster and therefore ideal for percussive instruments or any instrument with a fast leading edge to its sound. Larger diaphragms tend not to be quite so responsive, but are ideal for vocals and choirs. Then there is the polar response to consider and the microphone self-noise, which is how much noise the circuitry within the microphone produces. By the time the sound wave has been picked up by the microphone and sent down the microphone cable (which has properties of its own which will affect the sound) into the mixer's input socket, it has already been subject to a certain amount of degradation. Having been pushed around the mixer and, finally, sent to the tape recorder, it will have suffered further degradation. Then, the tape recorder itself may add a little noise or distortion. These amounts of signal degradation are very small, but they are cumulative.

By the time most studios were equipped with 24-track recorders and suitable mixers, microphone design had become well established, and there was a huge choice available, from affordable to ridiculous in cost and, as with so many other things in the world of electronics, one quickly came up against the law of diminishing returns. These days, quite good microphones are readily available and affordable. Consistency is sometimes an issue with lower cost designs, but they are still useful. In the tape world, having got to the point of producing 32-track monsters with 2-in. or even 3-in. tape, digital audio came along. And so, at first, they produced digital 32-track monsters, still running tape. These were expensive and complicated machines. Nowadays, things are very different indeed and most recordings are made, via an interface to computers which may be custom built devices or off the shelf designs. There are also some hybrid designs that record to either solid state media or hard discs which then need to be connected to a computer or a digital mixer, for subsequent mixing and production. We have come a long way since the BTR machines showed what was possible with tape. Many, including the author, miss these tape machines. There was something very satisfying about getting a good recording on tape, and the tactile factor of threading tape through its path and checking levels was all part of the enjoyment. It also represented a wonderful learning path for future audio engineers to follow.

Chapter 6
The Development of Microphone Techniques

Blumlein had defined microphone techniques for stereophonic sound in 1931. His idea of a coincident pair of microphones at 90° to each other, with the capsules aligned vertically, works very effectively, even today. This configuration produces a wonderful three-dimensional sound field which is highly compatible with monophonic sound. Consequently, early stereo recordings could be broadcast via monophonic radio channels if required. This 'coincident pair' as it is referred to may be used with microphones of either a cardioid or a figure of eight polar response. In practice, it is mainly used with cardioid microphones. The 'Blumlein coincident pair' as this technique has often been referred to has been used on countless recordings and live broadcasts and is as valid today as it was in 1931.

Inevitably, record companies and, later, broadcasters began to experiment with similar configurations in order to capture a sound which was particular to their organisation. In the 1960s and 1970s, the 'Decca Tree' was widely used for classical recordings. It consisted of a special framework which was highly configurable but consisted usually of a coincident or near coincident pair with a third microphone placed above or behind the others. Sometimes even four microphones were used in various arrays. This provided Decca with the flexibility to go out to a remote concert hall and quickly obtain a microphone configuration which worked well both for the venue and the piece of music being recorded. For example, an orchestral piece which also featured a choir would use a particular configuration. An opera might use a subtly different configuration. The Decca engineers and producers were highly experienced and could be relied upon to produce the right sound. Consequently, Decca classical recordings made from even the late 1950s to the early 1970s sounded fabulous, and many are held up as landmark recordings, even today. Decca understood stereo and took the concept even further.

The Mercury Living Presence label in America, thrived with the influence of freelance engineer Bob Fine and in-house coordinator/engineer Wilma Cozart. They also had the idea of going out to remote concert halls, but they pioneered the use of a location recorder, in their case a medium sized van with recording equipment in it.

© The Author(s), under exclusive license to Springer Nature Switzerland AG 2021   25
J. Ashbourn, *Audio Technology, Music, and Media*,
https://doi.org/10.1007/978-3-030-62429-3_6

They typically used three microphones in an array in front of the musicians and mixed the resultant signals down to two tracks via a simple level only mixer. They experimented with various tape formats, which had to be reasonably portable, and even with recording three tracks onto 35-mm film stock. Their simple approach ensured that recordings sounded fresh and 'alive' and their willingness to travel overseas, including to Russia, ensured that they captured some unique performances. The Mercury Living Presence label quickly became associated with high fidelity and was very popular. These early recordings by Fine and Cozart, from the 1950s to the early 1970s, remain popular today and have been reissued on CD format. The recordings they made were not strictly stereo, but they were impressive and did contain some spatial information when played back on a stereophonic system.

The so-called AB microphone technique is sometimes used in live recording. It consists of two omni-directional microphones placed in line on a bar about 8–16 in. apart, depending upon how close they are to the source. Sometimes, this technique is used to record a single instrument, such as a piano, where it can provide a sense of space without dividing the piano into two for left and right channels. It can also work in an orchestral context, but one needs to be careful of phase anomalies due to the position of the microphones and the distance from the source. There are several other options available to the recording engineer and one wonders why anyone would bother with this particular technique, but it can work well if properly configured.

The coincident pair configuration, already mentioned, is sometimes called the 'XY' technique. It involves the use of two cardioid microphones placed so that their capsules are in vertical alignment, one above the other. Specialist mounting brackets are available in order to achieve this. This technique will provide a proper stereophonic, three dimensional field and works well for most situations. It is also very compatible for monophonic use, such as being broadcast on a monophonic radio channel. The only real criticism of this technique is the width of the stereo image that it provides. If recording a large orchestra which is spread across a broad stage, then one of the near coincident techniques may provide for a wider stereo image.

Some of the national broadcast stations have undertaken their own tests and come up with variations on the near coincident technique. One of these is the French ORTF configuration which uses two cardioid microphones, spaced 7 in. apart and angled at 110°. This gives a nice, wide stereophonic spread and is particularly suitable for recording orchestras. Some stereo microphones are available, purpose built into an aluminium housing that adheres to the ORTF standard. This makes it very quick and easy for the audio engineer to set up and record without difficulty. It also helps the equipment used to remain unobtrusive, and this is sometimes very important.

The Germans have a DIN standard which specifies two cardioid microphones spaced 7.8 in. apart and angled at 90°. This effectively modifies the coincident pair standard to in order to separate the capsules from vertical alignment and have them pointing outward at the specified width. This provides a stereophonic image slightly narrower than ORTF but still a little wider than the traditional coincident pair. It is a good compromise and works well in most situations.

In the Netherlands, a system named NOS represents yet another variation of the near coincident pair. In this case, the cardioid microphones are spaced 11.8 in. apart and angled at 90°. The author has used this technique on occasion, to record orchestras which are not optimally deployed in the spatial domain. For example, when a large orchestra is trying to fit around a small stage and, consequently, has some 'floating' musicians who are not in their usual position.

All of the above techniques use just two microphones in order to capture a stereophonic image. Broadcasters warmed to these simple deployments because they could easily be set up very quickly, and quite often, there was little available time to do this before a live broadcast. It is curious (and quite wonderful) that some of these broadcasts which were also recorded are now available on CD. They may often be found on obscure labels and in low-cost box sets, but they are a real treasure, partly because of the proper stereo sound and partly for the performances which are usually excellent, especially if they have come from Poland, Russia or Hungary.

In recent years interest in Ambisonics, which was first developed in the 1970s, has arisen, mostly because of the virtual reality industry which includes computer games and films. Ambisonics is a surround sound system which is highly configurable. In its most simple, popular form, Ambisonics B format, it utilises four microphone capsules within a single housing which is placed at the centre of the required sound field. The four channels are usually referred to as W, X, Y and Z. In simple terms, W has an omni-directional polar pattern, picking up all the sounds within a 360° sphere. X is of a figure of eight polar pattern, pointing forwards. Y is of a figure of eight polar pattern pointing to the left. Z is of a figure of eight polar pattern pointing upward. Together, they provide all the spatial information necessary for a full 360° sound field with height and depth information as well. The clever thing about Ambisonics is the software. Within software an audio engineer may direct a signal in almost any direction as well as controlling movement. Playback of an Ambisonics signal is speaker agnostic. It may be configured to play back on headphones, on a two-speaker system, a four-speaker system or more. This basic Ambisonics B format is noted as a first-order system. Second-order Ambisonics uses nine channels and third-order Ambisonics uses 16 channels. The most flexible of all is sixth-order Ambisonics which utilises 49 channels! It would be a strange project which required sixth-order Ambisonics. But the popular first-order Ambisonics B, with four channels can provide interesting opportunities for the inventive audio engineer.

Hence, microphone technique has evolved over the years, and yet, with the exception of Ambisonics, most configurations have been an adaptation of Alan Blumlein's original invention of 1931, and his suggestions work perfectly well today. As a part time audio engineer, the author has experimented with many configurations and, for most cases, has used either a coincident or near coincident pair as the main microphones with, occasionally, two more microphones configured as a separate array if, for example, a choir is present and is situated well behind the orchestra. The signal from these two microphones will then be mixed in, usually in small amounts, to the main signal in post processing. However, on many occasions, the use of a single pair of cardioid microphones has proved to be sufficient.

A technique used by the author, and no doubt other freelance audio engineers, is to visualise the sound field when setting up the microphones. Knowing the polar pattern of the microphones being used, one may look at the direction in which the microphone is pointing (and vertical angle) and visualise the sound field that will be captured. With practice, this technique works very well and may or may not suggest the use of a second pair of microphones in order to cover every eventuality. It certainly does no harm to have a second, stereophonic recording at hand during post production. This may then be blended, in a small amount, with the output from the primary stereo pair, if indeed required. Often, it has been found not to be required, but it is nice to have it available (Fig. 6.1).

Fig. 6.1 A large diaphragm condenser microphone

Chapter 7
Multi-Channel Tape Recorders

Tape recorders have already been covered to a certain extent, but here we focus particularly upon multi-channel tape recorders and the implications of using them to a professional standard.

With a basic two-track or four-track tape recorder using quarter-inch tape, aligning the heads properly is not too difficult. On a three-head tape recorder (erase, record and play), it is necessary that each head is properly aligned in both the vertical (zenith) and the horizontal planes (azimuth). In the vertical plane, it is important that the tape exerts equal pressure across the surface of the head. If it does not, then drop-outs may occur or one channel may sound weak. On record *and* playback, this effect is doubled. In the horizontal plane, if the heads are not properly aligned, the track will not be fully recorded on or fully played back. As you can imagine, if a tape is recorded on a poorly aligned machine, it will play back poorly on a correctly aligned machine. It is the engineers' job to ensure that tape recorders are properly aligned, and they do this using a special tape of test tones which have been recorded by a third-party organisation.

In the early days, 'tape men' were separate from the audio engineer and producer, and they would take great care, and pride, in ensuring that the machines under their jurisdiction were properly aligned. Later, the in-house audio engineer at a recording studio would be expected to keep an eye on such things. However, this was not always the case.

If this alignment is critical on a two-track machine, then, as you can imagine, it is even more critical on a four-track machine using the same tape width. The Fostex company even made an eight-track machine (the R8) that used quarter-inch tape and was targeted at amateur home studios. Other eight-track machines appeared with half-inch tape and Ampex made a proper, professional eight-track machine using 1-in. tape and separate preamplifier boards for each channel, designated the AG440 8-A. The Brenell company offered the type 19 1-in. eight track, also with separate preamplifier boards, and the company was subsequently bought by the mixer manufacturing company Allen and Heath. The new organisation produced the much loved

J. Ashbourn, *Audio Technology, Music, and Media*,
https://doi.org/10.1007/978-3-030-62429-3_7

Brenell Mini 8, which was a professional machine at a cost within reach of amateurs and musicians. No doubt, many popular songs originated on a Brenell Mini 8, they were wonderful machines. There were also machines available from 3M, Tascam and others. Aligning the heads on 1-in. eight-track machines was critical, but not too difficult.

But then 16-track machines came along, using both 1-in. and 2-in. tapes. And then 24 track on 2-in. tape. Each of these advances in technology rendered head alignment even more critical.

If head alignment was an exacting task, so was the alignment of the electronics, which had to be precisely undertaken for each track. If noise reduction systems, such as Dolby A or DBX were used, then these, which would typically be on separate cards per track, would also need to be precisely aligned. This is also achieved with test tones at certain frequencies but is useless to perform if the heads have not been properly aligned first. It goes without saying that the heads also need to be regularly cleaned and demagnetised. The former may be achieved with pure alcohol from the chemist, the latter with a special, but readily available tool.

If these adjustments are not performed regularly, some recording sessions will yield better quality than others. It was not unusual in the late 1960s to early 1980s to find that recordings, especially those in the field of popular music, varied quite dramatically in sound quality, and even classical recordings were not immune from this effect which may be directly related to how well multi-track recorders were being maintained. The next variable would, of course, be the mixing process down to a two-track master. Each recording studio would have its own acoustics and so would each control room, where this mixing would be undertaken. Furthermore, audio engineers tended to have their own preferences around playback monitors (loudspeakers). In some cases, they had no choice as the monitors would have been custom built into the control room. Nevertheless, different monitors had different tonal characteristics, further affected by the design of the control room itself. In addition, the amplifiers used to drive them would have their own particular aural characteristics. A combination of bright sounding amplifier with bright sounding monitors would usually result in a dull sounding recording, as engineers naturally compensated for the bright sound.

Of course, these variables would not be so pronounced in a well-designed studio and control room, where the tape machines were well maintained. In such a case, the resultant two-track master would sound excellent and be ready to be used to make one or more disc stampers. For disc playback using a magnetic or moving coil playback pickup cartridge, there was a further complication in that something called an RIAA (Recording Industry Association of America) equalisation curve was encoded into the record, to enable a smoother and wider range playback while also permitting longer playback times from the finer gauge grooves. The problem being that not all recordings got the time constants for this right when making the disc stampers. Consequently, some records sounded unnaturally bright and some sounded unnaturally soft. Fortunately, although these were isolated cases, I expect that most record collectors found examples among their collections. In addition, the disc stampers deteriorated with age, resulting in manufactured records gradually becoming of poorer quality. Ironically, the more popular the recording, the more likely it was to suffer from this stamper deterioration.

And so, as we can see, the path between the original microphones and tape machines and the finished disc was fraught with dangers to the fidelity with the original sound. Mostly, it worked quite well, but there were significant variances, and these were often due to the tape recorder being used.

There is another aspect of multi-track recording which deserves mention. Musicians liked multi-track because it enabled them to add endless overdubs to their original tracks. The problem is, every time the tape is pulled across the heads, it loses a little of its oxide. With each pass, the sound becomes duller, losing both transients and high frequencies. The effect is small but cumulative, so once the musician is on his tenth overdub, the sound quality is suffering. Multiply this by the number of musicians, and it can become an issue.

While we are on the subject of tapes, there were of course a number of companies manufacturing tape. 3M, Agfa, BASF, TDK, Ampex and others. While based on similar principles, the fact is that they all sounded slightly different and would have different electrical bias requirements. This is another reason why the multi-track recorder needed to be carefully aligned to the actual tape being used. In addition, there would often be small differences between batches of the same brand and type. Imagine also that a tape which has been recorded in one studio, is transferred to another studio for further overdubs, by no means an unusual practice. Ideally, the multi-track recorder should be realigned to this particular tape, but how often would this be the case? Hardly ever would seem to be the answer. And yet, brands such as Agfa and TDK were quite different in their tape formulations and subsequent performance.

Another issue was tape compression. Audio engineers tended to have different views in this respect and much would also depend upon the mixers being used and their level metering. If both the mixer and the tape recorder featured fast, peak level metering, then they could achieve a reasonable control over tape saturation. However, many mixers provided the option of using slower to respond VU (volume unit) meters, which some engineers preferred as they were easier to look at over long time periods. The problem with VU meters is that they did not register peaks in audio levels, so a level set roughly at 0VU was bound to over saturate the tape, leading to distortion. Now, there are plenty who will claim that tape distortion (mostly third harmonic) is not necessarily unpleasant to listen to. Not everyone will agree on this point. But most will agree that the compression effects caused by tape saturation *are* unpleasant to hear. Some popular musicians claimed to like tape compression as it made their songs sound louder. Louder, yes, but also distorted.

It is not widely appreciated that the average early 24-track tape recorder actually had poorer specifications in terms of tape noise and high-frequency response than some of the up-market hi-fi reel to reel machines from companies such as Akai, Pioneer and Technics. However, the noise issue was ameliorated by the widespread use of Dolby A, and especially with popular music, there was little high-frequency content beyond the range of these machines. Another point to consider is that most tape machines gave their flattest frequency response at a level corresponding to a true

0VU. So there was really no need to push these machines into tape saturation, but they frequently were pushed. For classical recordings, which have a wider dynamic range, most pieces would be accommodated within the noise floor of these machines but, in any case, the audio engineer could compensate by temporarily reducing the level on known loud crescendos (which he would know because he would have been present at rehearsal and will probably have the score in front of him).

All of these variables explain why discs mastered from tapes featured a wide range of perceived quality. As mentioned disc stampers also deteriorated as they aged, affecting sound quality, but this was rarely too much of a problem as reputable companies would ensure that they were replaced after so many cycles. Recordings made on tape had a distinctive sound which many came to love and which many still miss. There was something truly organic about this sound, and engineers, of course, loved to play with these beautiful machines, some of which were engineering works of art, providing they were properly aligned (Fig. 7.1).

Fig. 7.1 Tape head alignment

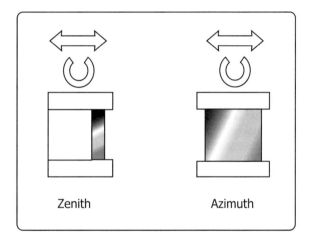

Chapter 8
The Advent of the Big Studios

Big, in this sense, may be defined in two ways, the size of the performing area and the size and equipment level to be found in the control room. Originally, before the boom in record sales and the advent of the tape recorder, most musical events were broadcast live over the radio. This meant that broadcast engineers went to the place of performance, usually a purpose-built concert hall or, occasionally a church, set up their microphones and connected back to the broadcast studio by land lines. Most broadcasters had small studios suitable for interviews and plays, but could not house full-scale orchestras.

With tape recorders and a huge demand for records, it became obvious to the record companies that they needed their own studios. In general, these purpose-built studios, which had to be sound proofed and were therefore expensive, started out fairly small, to house the popular singing groups and jazz bands. For classical music recordings, most labels still went to the concert house (as many do today) and set up their equipment, with tape recorders in a separate little annexed room, often called 'the green room'. The only downside with this approach was that the audio engineers usually just had the one chance to record a single performance. If anything went wrong, it *really* went wrong. Even when taking two recorders, as was the usual practice, if the performance was sub-standard, then a good recording would not result.

There was also the advent of recording sophisticated orchestral soundtracks for film. Of course, an existing concert hall could be hired, but this was an expensive business, and you still had to get all the recording gear there. The administration alone would be a major undertaking. And so, recording studios started to evolve to become larger, more sophisticated and offer a wide range of recording facilities in purpose-built recording and control rooms. Naturally, some would be better executed than others with the poorer designs suffering from noise breakthrough and bad ventilation. On the other hand, some started out the way they meant to continue.

The world's first purpose-built studio was built in London by The Gramophone Company in 1928. It opened in November 1931, by which time The Gramophone

J. Ashbourn, *Audio Technology, Music, and Media*,
https://doi.org/10.1007/978-3-030-62429-3_8

Company had become EMI. It was the same year that Blumlein invented stereo-phonic sound. It was, and remains, the world's largest recording space. Today there are three studios plus other facilities, and its address is Abbey Road, St John's Wood, London. Today, Studio One can accommodate almost any recording project that you could imagine, and of course, by now, they have accumulated a vast array of classic recording hardware.

One of the popular audio engineers/producers at EMI was George Martin who famously signed and recorded the Beatles up until 1970. He eventually cobbled together a group of like-minded engineers and producers and set up AIR studios which opened at Oxford Circus in October 1970. In 1979, after the lease had expired at Oxford Circus, George Martin opened AIR Studios Montserrat, a small island in the eastern Caribbean. In 1989, the studio was hit by a hurricane and 90% of its structures were damaged beyond repair. George returned to London and acquired Lyndhurst Hall in Hampstead. It opened in December 1992 and continues under new management today. It is a magnificent building, originally constructed in 1884 and has wonderful acoustics for recording.

Such are the twists and turns of fate which accompany many of the large studios. Abbey Road continues and is so well known that everybody wants to record there. Some are not so lucky, and unfortunately, many big studios around the world have had chequered histories and have fallen by the wayside, while others continue to open, like Mark Knopfler's British Grove Studios in Chiswick. This is a custom-built studio which is very quiet and very 'clean' electrically. It features a mixture of old and new equipment and will doubtless prove popular, partly due to its conve-nient location. Sadly, British Grove Studios lost one of its most influential team members, Dave Stewart who passed away in April 2020.

Probably the most consistently successful studios around the world are the inde-pendent medium-sized affairs which can cater for popular bands and musicians, jazz combos and others of the sort. Those who can house full orchestras comfort-ably will always be in demand, but the cost of equipping and running such a facility is by no means small. On the other hand, studio rates have shot up quite consider-ably since Abbey Road first opened its doors. The other cost which people may underestimate is proper studio acoustic design. Ideally, in a large studio, the acous-tics should be fairly neutral with no nodes of excitement and a known, gentle rever-beration time. This is why in many such studios, you can find 'bass traps' on the wall to counter standing waves and areas of absorption to tame the high-frequency reflections. There was a trend, largely in the 1970s, to design studios with a 'live end' and a 'dead end' in order to cater for every eventuality, but this remains a com-promise. These days, designers can plot the acoustics on a computer using impulse tones at precise points and measuring the response. This enables them to identify any problem areas and deal with them accordingly.

It should be noted that with the advent of the Digital Audio Workstation (DAW), many forms of music may be made on a computer with a digital audio interface. This means that such music may be made, in theory, almost anywhere, even in the aspiring teenager's bedroom. This is true enough, but such music will still require mastering to accepted standards and few will have the skills, or equipment, to do this. However, it does create the possibility of building a small studio and mastering suite based upon the use of computers and digital music. This may be ideal for composers of film music, for example at least to enable the creation of good quality demonstration material. It may be that full orchestras and large studios are still used for the final sound track.

We live in interesting times. At first, there were few studios and most that did exist were reasonably large. Then came the 1960s and 1970s and a crop of medium-sized studios shot up in order to cater for the burgeoning popular music industry. Many of these became extended as they all wanted 24-track facilities, and many are now converted to run computers. What on Earth happened to all those beautiful 8-, 16- and 24-track tape recorders? For a while, they could have been bought at bargain prices, but now they are collectors items. Similarly, the original, big 72-track mixers must still be hiding away somewhere or may have been turned to scrap. But new models are still being built and offer a wealth of customisable features to those who can afford them.

In conclusion, the 'Big Studios' are still with us, but many have changed their complexion somewhat, and sadly, many have disappeared. There are still many great recordings made on location, and in these days, the audio engineer can transport everything required in a large domestic car! Some famous engineers have been known to tack everything they really need in a rucksack and use existing cables and microphone stands or simply hire them in. How things have changed. And for the performers as well. The awe that one must have felt upon entering your first 'proper' studio, and seeing all those wonderful toys in the control room, as well as being lost in space, until everyone else had arrived. What fun they must have had in the late 1950s and throughout the 1960s making that huge legacy of recorded music which we have today. By the 1970s and 1980s, it was all getting a bit cut-throat and commercial, but there were still those around who remembered the good old days and the tricks employed to get a good sound down on tape. Now, everything is digital, and there are, of course, pros and cons to this approach, but most of the old techniques still work. The question is, who is around to remember them and pass them on? And that, indeed, is a good question.

Chapter 9
The Record Business

When discs and gramophones became popular, it was evident that there was money to be made in the record business, especially in America, where the number of units sold, for any popular recording, could be massive.

I remember there was a scene in the film made about the life of Glenn Miller, starring James Stewart, where the father asks Miller how much he gets paid for one disc. I think the reply was 'three cents' the father reflects, 'that doesn't seem very much for all the effort' then the father asks 'how many do you sell'. When Miller tells him, he quickly works out the sum in his head and then beams 'that's a good business isn't it?' Recording artists would also receive a royalty each time their record was played over the radio. The percentages were small, but the scale ensured that many of them became very wealthy indeed.

With the explosion of popular music in the 1960s, young men and women suddenly found themselves earning more than top executives in industry. Unfortunately, some of them did not know how to manage their new found wealth and turned to alcohol or substance abuse. This seemed particularly prevalent in the jazz scene, both pre-war and post-war. Others were more sensible and enjoyed sustained careers, once they had crossed a certain threshold of popularity.

If the recording artists earned a small percentage of the revenue from discs, the recording labels, even after the costs of manufacture and distribution, earned a large percentage. In the popular music field, this enabled them to take chances with signing new artists and hoping that they would prove popular. In the 1960s, in Britain and Europe also, it seemed that no one could go wrong, and vast fortunes were made. Occasionally, of course, things did go wrong and everything came crashing down around them, but this was ameliorated by the fact that many popular recording labels were simply outshoots or imprints of the big labels and could be created or destroyed at will.

The classical music scene was rather different. In this case, individuals would take great care in building their own library of classical music. They would go to concerts and note which orchestras were especially good, listen to the radio for certain recordings or performances and read the reviews in the dedicated magazines. Consequently, there was always a steady demand, with less peaks and troughs than in popular music. Furthermore, classical music fans would often buy two or three versions of their favourite pieces. The more dedicated classical music consumers would (like the author) build massive libraries of the outputs of almost all of the classical composers. These days, the classical music labels have another bonus in that the business has been in place long enough for them to reissue recordings for which the copyrights have expired. This enables an even bigger cut of the profits for themselves as they are also not paying for the costs of recording. This is also happening now in the jazz world, with great success for those involved. In addition, once they have established distribution, which may be largely online, the cost of entry into this market is minimal. They can transcribe from the original discs, clean up the sound in software and issue the recordings to eager buyers. And eager buyers abound, because many realise that the so-called golden age for jazz really did produce some of the best performances on record. Similarly, in the field of classical music, many of these older recordings, some of them were simply radio broadcasts which happened to have been recorded, featured peerless performances by some of the world's greatest musicians and conductors.

The record business has itself seen many changes, with both down-slides and new opportunities. The advent of the Compact Disc was a boon for record companies across all genres as it enabled the reissue of virtually their entire catalogue, at minimal cost. Unfortunately, some of the transcriptions to CD have been of poor quality, but recordings that were only ever made for CD have fared rather better in this respect as both engineers and producers have come to grips with the digital world.

In the field of classical music, there are literally hundreds of record labels. There are the big labels that everyone remembers such as EMI, Decca, Deutsche Grammophon, CBC Records, Philips, RCA, Sony, Warner and others. Then there are labels which are considered as more specialist, for one reason or another, such as Harmonia Mundi, Telarc, Accord, Brilliant Classics, Denon, Hyperion, Naxos and many others. The big labels can mount large-scale productions, hiring the best studios and bringing together the best orchestras and conductors. Moreover, they can allow them the time to settle in and rehearse properly, consequently getting consistently good performances. Exactly how they go about the recording is another matter, which shall be explored later in this work. The smaller, more specialist labels may record at live events or may bring together smaller choirs and chamber ensembles to record at venues which are not recording studios, such as churches. Some of them, like Naxos, have always encouraged new artists and have kept down the cost of production and distribution. Others have specialised in particular genres such as Early Music, or little known pieces performed by equally little known ensembles. This works quite well as, often, it is these smaller labels who will go to great lengths to capture a good quality recording. The author has many such discs, and they are among the most enjoyable to listen to.

In the popular music field, there was, in the 1980s and 1990s an, explosion in what were known as the independent labels or 'indies' as they were usually called.

These were often very small concerns, started by enthusiasts or, sometimes, the musicians themselves. Some of them found instant success and grew to rival the established labels. Others hit a peak and then collapsed or were absorbed by the larger competition. But they were an important development for young artists who could not get signed by any of the established labels, but found a ready welcome with the independents. They took big chances but were very often rewarded for their commercial courage. In addition, there were many young engineers and producers who started their careers with the independent labels and have subsequently moved on to greater things. In general terms, they were a great success and brought several new artists to the public attention who might have otherwise remained unheard of.

Nowadays, for popular music, the tide has turned again and aspiring artists can find many ways of getting their music online for free. In addition, there are 'virtual labels' and companies claiming to get your music in front of the artist and repertoire executives of the big labels. Similarly for film music, there are companies who claim to successfully field your music. It is also possible to use the services of 'manufacture on demand' companies, which is another way of getting music onto the Internet and, of course, young artists can make full use of the social media sites in order to promote their music. Unfortunately, this glut of opportunity is both a blessing and a curse. On the one hand, yes, young artists can get their music available online. On the other hand, there are millions of them, all over the world, doing just this. Consequently, finding anything that appeals to you may be a lengthy business as there is no quality control whatsoever, so anyone can put up anything, especially if they are prepared to pay one of these 'promotional' companies that are also springing up everywhere. It is they, if anyone, who will benefit from this approach. Nonetheless, there have been one or two lucky individuals who have come through this channel. And this is the problem. It reduces their chances to pure luck, whereas if they had been signed by a major label, the latter will have invested in, firstly, a good quality recording and, secondly, in proper promotion, ensuring that they had a good chance of becoming known to a wider public. One may take a similar approach with classical music, the problem being, why should anyone buy from an unknown when they may purchase confidently from an established label?

One attempt at crossing these hurdles has been made by the many 'streaming audio' companies that have sprung up and continue to appear online. For a small subscription, the consumer has access to a wide range of music which they can listen to over the Internet or even download and keep on their own computer or smart phone. There are many issues with this approach. Firstly, for high-fidelity enthusiasts, some of them offer 'high-resolution' streaming or download for use on high-quality audio players or high-fidelity components. The problem here is that many such recordings have simply been up-sampled from the original CDs. This is certainly not true high resolution and yet requires an enormous bandwidth to download and use. Secondly, many encourage the use of compressed formats such as MP3 which, for use on smartphones, is probably acceptable, but is certainly not high fidelity. In addition, because the user is exposed to so much choice, the novelty often wears off rather quickly and they revert to purchasing the music which they know from the labels which they trust. In general terms, the ready availability of music, tested and untested, via the Internet is more of a problem than an advantage. Some may disagree, but there is no substitute for properly performed and properly produced and recorded music by established artists. The Internet blurs this distinction somewhat, especially for those who are inexperienced in understanding good musical performance and high-quality production. Nevertheless, this seems to be the way that a majority of people are consuming music, probably varying somewhat by country.

In conclusion, the record business has come a long way, and from the consumer's perspective, there remain treasures to find, including new releases on vinyl discs. But the traditional way of producing music has become somewhat endangered by these developments. Just how this will mutate in the future is anybody's guess, as things do seem to be in a state of flux at present. Let us hope that this settles into well-defined channels that we can trust over time.

Chapter 10
The Maverick Producers and Freelance Engineers

In the early years of recording, the big record companies tended to have their own in-house engineers and producers, who were very carefully trained, not only in their respective crafts, but in the particular style and way of working of the company. This was also true of broadcasters such as the BBC who maintained extensive training facilities and rigorous methodologies, ensuring that a BBC engineer, by the time he was deployed, was already highly skilled. Indeed, the BBC maintained their own equipment engineering team who produced many of the world's finest broadcast monitors, for example. And so, this was the accepted way of doing things. Decca Records, throughout the 1950s and 1960s, produced some wonderful classical music recordings. Sometimes, these would be recorded in their own facilities; often, they would be recorded at venues throughout the world, where Decca engineers would be sent, together with all their recording equipment. They would set up on site, usually the day before rehearsals and then check and double-check everything, from the quality of the AC supply, to every cable and microphone. Very often, the conductor or orchestra leader would have their own ideas about the sound they wanted to achieve, and they would be invited into the control room to discuss accordingly. Sir George Solti was one such conductor who knew his way around the control room. He was particularly keen on opera and held many prestigious positions around the world during his career, including with the Covent Garden Opera Company in London.

The engineers and producers of the main labels worked closely with the musicians and conductors and, yet, maintained their own ideas about technical excellence. This is what lead to the famous 'Decca Tree' and the wonderful recordings they made with this particular microphone configuration. Other labels had their own techniques that they liked to follow, as well as technical preferences when it came to microphones, mixers and tape recorders. This is why original recordings made by these big labels tended to have their own sonic signature. It was a subtle one, often playing to the conductor's wishes, but it was there. Record buyers came to understand this and would have their own favourites, although it would not stop them

J. Ashbourn, *Audio Technology, Music, and Media*,
https://doi.org/10.1007/978-3-030-62429-3_10

from buying from other labels if the performance was noteworthy. In Britain, *The Gramophone* magazine was established in 1923 and, ever since, has become regarded as an authoritative source of record reviews. In the 1950s and 1960s particularly, classical record buyers looked to *The Gramophone* for guidance in building their own libraries and those recordings which won a Gramophone award could be relied upon to be especially good.

Such was the status quo for a long time. And then, slowly, independent audio engineers and producers started to spring up. Sometimes, they had simply left the big companies and started out on their own. Often it was because they could not find a position in one of the established companies that suited them. From the record company's perspective, this was quite useful because, if they found themselves overloaded with work at their own facilities but still needed someone to cover a special event, they could simply hire one of these 'maverick' freelancers. Of course, they took a chance in doing so as any resultant recording, if used, would carry their name and reputation. Consequently, in the world of classical music especially, these independent engineers and producers had to be good.

And some of them were very good and quickly became recognised for the quality of their work. One such freelance engineer who has recorded works for all the main labels and several specialist labels is Tony Faulkner. Tony has recorded literally thousands of pieces and always attains an exceptional level of quality. He was one of the first to successfully use digital technology and, combined with his skillful microphone technique, managed to make early digital sound good, while others were struggling with it. I have several recordings that were engineered by Tony Faulkner, and they are all marvellous listening experiences. It is interesting to learn that Tony always had a passion for music and was introduced to the world of recording when his father bought him an early Telefunken tape recorder. At University, Tony was already working with others on 360 degree immersive audio, many years ahead of what is being done now, and quickly went on to work with leading conductors and orchestras around the world. Tony Faulkner's example is an important one as it demonstrates so clearly that one must have a passion for what you are doing, if you are to do it well. It is also most refreshing that he has always been most considerate of the musicians themselves, keeping in the background as much as possible and letting their art come through. Perhaps that is why so many of his recordings have won awards. Interestingly, one of Tony's often used microphone techniques, the phased array of four microphones on a bar, the outer pair being omni-directional, has its origins in RADAR, which brings us nicely back to Alan Blumlein.

Others also created a niche for themselves, and in the field of popular music, there have been dozens of them. Unfortunately, these days, everyone seems to consider themselves an audio engineer, whether or not they have the requisite training and background, and this sometimes comes through to the finished article.

A good freelance engineer enjoys the freedom to make decisions on the spot, especially when encountering the unexpected, and also, naturally, takes the responsibility for those decisions. As an example, the author was once committed to recording an orchestra playing favourite film scores. Knowing the theatre being used, a microphone placement approach was already in mind, when entering the theatre, the orchestra was found not on stage, but crammed into the space between the stage and the front row of seats. This was because excerpts from the films concerned were also being shown on a large on-stage screen, the orchestra playing in synchronisation with what was being shown. A nice idea, but it left nowhere for the audio engineer to set up his recording equipment. In the end, the author was wedged into the second row of seats with a single microphone stand, precariously placed between the first two rows, upon which were placed an array of four identical cardioid microphones, the inner pair roughly corresponding to the NOS specification and the outer pair roughly corresponding to the ORTF specification. The resultant spread of acceptance suited the unusually arranged orchestra. The output from the microphones was fed directly into a four-track digital recorder, with the limiters switched off and no mixing device being used. The resultant sound, when mastered, was crisp and clear and unusually good considering the duress under which it was gathered. The author had, in addition, to tactfully persuade the audience around him to keep quiet during the quieter passages. Such is the excitement of live recording.

On another occasion, the author was required, with just 1 day's notice, to record a large (and very loud) concert band. It was the band's 25th anniversary performance, and they wanted a nice memento of the occasion. With only about 1 h's rehearsal time being available to him, the author was in a rush to set up his microphones and find a suitable, unobtrusive place to record from. The venue was a large church with reasonable acoustics, and the author was finally positioned behind a stone pillar but in line of sight of the conductors (there were several at different times). A large brass ensemble, with percussion, is not particularly easy to record as some instruments are more strident and prominent than others. The woodwind section, for example, tends to get somewhat overshadowed. On this occasion, a near coincident pair of cardioid microphones were placed over the conductors head and pointing into the middle of the ensemble. These were the main sound source. However, two other microphones were placed on either side in order to capture, in particular, the clarinets and flutes. This was a judicious choice as one of the pieces played featured a clarinet solo, although the author had limited knowledge of the repertoire that was to be played. It turned out to be a very good recording, even with the short notice and unfamiliarity with the venue, the repertoire and the conductors.

Hereby lies the value of independent audio engineers. They must be able to sum up a situation quickly, have a suitable variety of equipment to hand and make more or less instant decisions about how to record the live event. Moreover, as the event in question is generally only going to occur once, they have to make a successful recording. Naturally, they will also always make a backup recording and bring the two back to their base for final mastering. Often, the author, when recording, would make one high-resolution recording and one at normal CD quality. The distinction will be discussed more fully later on in this work.

Ironically, some freelance engineers would be absorbed back into the mainstream recording industry, usually due to the quality of their work. However, there is a definite requirement for freelance engineers and producers who, together, often make the most unlikely of recording projects successful, providing memorable outputs accordingly. Those with less technical knowledge and experience will not be able to achieve this, and this is somewhat worrying about some of the so-called independent engineers and producers today. Some of them are familiar with a certain brand of software, but lack the basic skills in physics and the knowledge of how sound works. However, as much modern music is produced from sampled sounds entirely in the digital domain, this may present less of a problem, although one still needs to understand how sound is manipulated in the digital domain and how not to 'over-cook' things too much.

We live in a fast changing world wherein there is scope to use a variety of skills with respect to recording sound and music (of course, it is not just music, the author has also extensively recorded bird song and other natural sounds). It is hoped that the traditional skills shall prevail, at least in the field of classical music where one does need a familiarity of orchestras, how they work and how they sound. Each orchestra has a distinctive sound of its own within a particular acoustic space, and capturing that signature sound is an important aspect of recording. The very word 'record' suggests the accurate recording of an event in time and that should surely be our goal.

However, the freelance engineer these days has much more than two-channel audio to think about. He or she must also be familiar with video and synchronising audio to video. In the amateur world, this is usually achieved by recording a short burst of energy (a 'slate' tone) from the recorder to the video camera's sound track and then, in post-production editing software, aligning these two tones in order to then replace the video recorder's sound track with that of the dedicated audio recorder. There are two reasons for doing this: firstly, because the audio recorder will have much better quality sound characteristics and, secondly, because it allows for freedom of movement of microphones in order to follow spoken dialogue or action.

In the professional world, the same thing is achieved via the use of timecode. In this technique, one or other of the (possibly many) devices being used will become the master clock and will send timecode to every other device (sometimes in a serial, daisy chain fashion). Then, in post-production, the picture and audio may be very precisely aligned to very fine intervals of time. This is essential when, for example, dubbing a foreign language onto the video, or for closely synchronising a sound track to the picture. All these skills and more shall need to be learned by the freelance engineer, at the very least, so that he knows what others are talking about in this context.

The aspiring audio engineer must also learn about Ambisonics and the various other surround sound techniques and how to record and apply them and, of course, be familiar with the workings of digital technology. It is indeed a different world now (Fig. 10.1).

Fig. 10.1 Leading audio engineer Tony Faulkner

Chapter 11
The Big Time, with 24 Tracks Everywhere

To go back to the formation of purpose-built recording studios and the tape decks that they had employed. Perfectly good two-track stereo tape recorders had been joined by four-track and then eight-track and, for special projects, 16-track tape recorders, usually with 16 separate preamplifiers, level meters and various optional circuit boards that could be added to them. This was really overkill as far as recording classical music was concerned, and many freelance engineers continued to make superlative recordings straight to two-track tape.

However, in the field of popular music, the ability to overdub onto existing material via recording onto adjacent tracks and then mixing them afterwards became instantly popular. And so, the tape recorder manufacturers played leapfrog, each time coming up with yet more tracks and working with the tape manufacturers to provide ever wider widths of tape, which had grown from a quarter of an inch to a full 1 in. and even 2 in. Technically, this was all very impressive, but were 16 tracks really required in order to make sophisticated popular songs? Consider that the very popular music of the Beatles and the Beach Boys, in the 1960s, had been made on eight-track machines.

But, of course, for a studio to boast that it had 16-track recording available, it gave a certain market advantage, especially in the eyes of popular musicians who tended to love toys and equipment. They could also boast that they had used 16-track recorders for their song or album, although whether they had actually used all of the 16 tracks is another thing altogether. Nevertheless, 16-track was regarded, for a while, as the leading edge in studio recording, and some impressive machines were built and installed in the leading studios.

Tape head design had also been moving forwards, with finer profiles and finer gap technology. It was realised that, with the advent of 2 in. tape, you could actually squeeze 24 tracks onto it without having crosstalk problems between tracks. And so, one by one, the 24-track monsters started to appear, from manufacturers such as Ampex, Studer, Tascam and several others. This allowed for all manner of experimentation and soon, all those studios who had rushed to install 16-track machines now had to repeat the process for 24 tracks. Did anyone really need 24 tracks? The sensible answer is no they did not, but, especially in the popular music field, musicians loved to fill up all the tracks with double-tracked guitars, vocals and more. And then things seemed to settle down, and 24 tracks on 2-in. tape became the norm for professional recording studios. Making the change was no small undertaking because it usually involved also a new mixer, with at least 24 main channels, plus sub groups. In fact, many mixers at this time had 48 channels or even up to 72 channels, but these were monsters that required complex installation and setting up.

Having filled up all of your 24 tape tracks, the time then came to mix them all back to two tracks for commercial release. No one can realistically be expected to make fine adjustments to 24 channels in real time, and so sub-groups were developed in order to group together similar elements, such as drums for example and then control them with fewer faders. Often though, these sub-groups were used simply to provide a mix for headphones used by the musicians in the main studio. The answer to this conundrum came with automation. Mixer manufacturers skilfully developed automation systems that remembered the fader and pan control positions for each channel strip on the mixer, enabling many passes to be made, gradually perfecting the mix. Of course, with every pass of the tape over the heads of the machine, it lost a little of its fidelity, but this really only became noticeable in extreme cases.

As a primary studio owner, you had to invest in not only 24-track tape machines but also a suitably sized mixer and then mixers with automation. Popular musicians always want to use the latest technology, and so, studio owners were faced with tough decisions about investment and studio rates. Some managed to get the balance just right, but some went under and disappeared forever. We were firmly in the age of 24-track recording and nothing else would do. Those who survived also had to invest in a considerable selection of 'outboard' equipment, such as reverberation units, compressors, limiters, delay lines and all manner of special effects which could be inserted into the mixer's channel strip or set-up as a 'send and return' device whereby a proportion of the sound from any mixer channel could be sent to the device and returned to the mixer on dedicated channels. All of this equipment, plus a broad choice of microphones, was expected to be available in a modern studio environment.

As the 1960s became the 1970s, the sound of popular music changed considerably. On the one hand, one could argue that it sounded more sophisticated and on the other that it sounded less natural. Voices were compressed and often swimming in artificial reverberation, instruments lost their natural balance in relation to each other, and very often, the lower frequencies (bass) were accentuated in a manner which made the recordings sound muddy and lacking in definition. Of course, this was not always the case, and there were also some very good recordings made, by those who knew how to do so. But the opportunity to experiment with all this lovely equipment did not always lead to good results. By the time the two-track masters had been used to create record stampers, many recordings of this era came across as over-compressed and muddy in sound. A case of too many studio toys being available to audio engineers who did not appreciate the effect they were having on sound quality, plus, no doubt, a strong influence from the musicians who insisted upon their use.

In the classical music world, things were a little different, but not absolutely ideal. Now that audio engineers had all these tape tracks to play with, it enabled them to put up a barrage of microphones, almost one per instrument in some cases, and then mixing the sound after the event into a pan potted two-track master which, of course, was not proper stereo at all. However, it enabled them to bring up the sound level of instruments playing brief solos, or for the primary instrument within a concerto. This made it very clear for listeners in terms of what was going on in these solo lines, but it was not, of course, what the composer intended. The great composers knew well the sound of an orchestra and how to score for it. Even given that, in the days of Handel, Haydn, Mozart and Beethoven, stringed instruments used quieter gut strings and some of the woodwind instruments were subtly differ-ent. However, these composers knew what they were scoring for and understood exactly how the holistic sound of the orchestra would sound. No one was better at this than the wonderful Beethoven, who used the instruments of the orchestra like a painter uses his palette of colours. Beethoven would have known full well that a flute or recorder solo line would sound much quieter than a full string section, but this was all part of the natural dynamics of an orchestra. Furthermore, Beethoven, as with some of the other great composers, tended to leave explicit instructions on his scores as to how a specific piece should be played. When audio engineers inter-fere with this natural balance, even with the best of intentions, they do a disservice to the composer as the work is not being heard as it should be.

It is the same with opera. Opera, which especially, should be recorded in true stereo with the balance exactly as heard on stage. This is important as opera is visual and the juxtaposition of players on the set is an integral part of the performance. If microphones are placed in front of individual singers and they are consequently forced to remain static, it is not the same thing at all. Fortunately, one may still acquire some of the early EMI opera recordings, which really do have a sense of space and excitement to them. Even the monophonic recordings have a sense of depth to them, due to the phase relationships between players and orchestra. The author has a copy of La Giaconda featuring the first recorded performance of Maria Callas. It is monophonic and yet, there is a very real sense of space and depth which makes for a rewarding listening experience.

And so, all the big studios now had 24-track capability with some very impres-sive, giant mixers, complete with automation. They also had 19-in. racks full of a variety of outboard equipment that could be plumbed into the mixer as required, and it seemed that almost everyone congregated in the control room at one point or another. What this produced, in the classical domain, was squeaky clean recordings with every line of every instrument or voice clearly delineated and raised in level at the appropriate moment. This could sound impressive, but it was not the natural sound of an orchestra, as heard from the podium. In the field of popular music, there was a free for all, with a great deal of experimentation, some of which produced interesting results, and some of which less so. A particular type of hard-edged, highly compressed sound developed which, for some, robbed much of this music of any inherent interest. The argument for this approach was usually that records made this way sounded better, or at least, louder, when broadcast via the radio channels. That was a shame, but, of course, as with all technology, once it is readily available, people will tend to use it, for better or for worse. It is often a case of technology leading the performance, rather than the other way around.

Chapter 12
How the Technology Changed the Music

The change to 24-track and associated hardware has been noted, but there were other changes. In the early days of recorded audio, there tended to be fewer components in the signal path and those that were used were very carefully designed, often by broadcast corporations or record companies themselves. There were not too many external suppliers. But now, anyone who could wield a soldering iron seemed to be making products for use in studios, as this was a lucrative market where, if a product became fashionable, it could be used to establish a sizeable manufacturing concern. Consequently, across Europe and America especially, there were dozens of companies providing studio equipment, from external processors to mixers, to studio furniture. This enabled a wide range of experimentation, with many new special effects being possible as well as the creation of many new sounds. This was also the time that synthesisers were making their way into studios, also using various external processors, with much connection taking place.

In the field of popular music, there exists a clear graduation across these time and technology boundaries. In the 1950s, we had the crooners, with simply recorded ballads and tuneful songs which others liked to sing in their bath tubs and, often, wherever they found themselves. If one listens to the songs, for example, written by Rodgers and Hart, it is surprising how many of them remain well known today. Throughout the 1960s, at first, the burgeoning rise of young, popular musicians also focused on relatively simple songs played live in the studio for recording purposes. But then came multi-track recording and things began to change. Popular bands such as the Beatles moved throughout this transition, and their later work became very different from the first songs that were heard from them. Into the 1970s, lots of tracks, lots of effects and, for some, lots of time in the studio to play with them. It was not unusual for the big name bands and musicians to spend several weeks in the studio, gradually building up layers of sound for their new record. This had a parallel effect on live concerts as, realising that the sound they had created on disc could not be replicated on stage, some bands simply became bigger, with more musicians

J. Ashbourn, *Audio Technology, Music, and Media*,
https://doi.org/10.1007/978-3-030-62429-3_12

crammed into the space and other visual effects, around lighting, for example being employed to make things interesting.

Staying in the field of popular music, together with its many offshoots, some of them with strange names, this trend has simply continued, with many sounds created in the studio which really cannot be easily recreated on stage. In fact, in many cases, especially with drummers, their playing simply triggers a pre-defined sound that has been created in the studio using sampling techniques. With synthesisers, it is possible to simply programme them with sequences, or even whole tunes, and then trigger these on stage. And so, in some cases, what is actually heard at a popular music concert is a combination of special effects, real and sampled sounds, with one or more individuals singing over the top. What one hears in the audience is certainly sound, but it is sometimes difficult to break it down into its constituent parts. Gone are the days when bands simply had two guitars, a bass, drums and a vocalist, at least for the most part.

There now also exist various sub-divisions of popular music, some of which rely upon specific rhythms, some on textures and atmospheres and some which are hard to categorise at all, but are a long way from the tuneful songs of the 1950s and 1960s. One reason for this is that anyone with a computer can make music quite easily. There exist several digital audio workstations (DAWs) which are available at no cost, and hundreds of sampled instruments or simply sampled sounds, that one may load into the DAW and, with the help of an external Musical Instrument Digital Interface (MIDI)-equipped keyboard, can create layered music, or at least sound, track by track. In fact, one does not even need the external keyboard as it is possible to just loop sounds together using a computer mouse and keyboard. The resulting sound tracks may be easily uploaded to any one of dozens of Internet sites from where they may be accessed. And so, the Internet is awash with quite literally millions of so-called songs from individuals who consider themselves to be musicians, whether or not they have ever studied music or even ever picked up a musical instrument. Those who have studied music properly and would like to enter into this field are thus somewhat disadvantaged because of the difficulty of ever getting their material heard over the mass of noise which now permeates cyberspace. No doubt they will revert back to live performance, if they can find venues who will accept them. The big names in popular music continue to make records in a more conventional manner and have them released by the equally big name record labels. However, even they are not immune from studio trickery.

The current situation within the popular music field is somewhat confused by this dichotomy between amateur and professional, the latter often made up by individuals who have worked through the past few decades. What we used to call 'Folk' music seems to have retreated into a specialist niche, as has the most traditional music based on geography. This is a shame, as there is much of interest buried within these niches. Jazz seems to have also experienced a similar, although less extreme, dichotomy, with those still playing what we recognise as jazz and others playing a strange mixture of sounds and styles which do not easily fall into any category, but are certainly not the sort of improvisational jazz that we knew from the

1930s through the 1950s and into the early 1960s. All of this has been made possible by the barrage of technology which is now available to almost anyone.

Within the field of classical music, the music itself has not changed, Mozart is still Mozart, but the way it is recorded has changed of course. In addition, we have what is often referred to as 'new classical' music, often pieces commissioned for a special occasion or just randomly, maybe by a broadcast corporation or even a record label. These pieces are interesting, but hugely variable. Some are written using traditional methods, employing traditional instruments and, consequently, sound as one might expect 'new classical' music to sound. Others do not use traditional instruments at all and are simply a range of sounds, often from percussive sources, sometimes synthesised, not arranged into anything recognisable as a structured piece of music. To the author's ear, this is not music at all, but some of these pieces do get broadcast over the airwaves and are critically reviewed. The problem here, of course, is that we simply no longer have people like Handel, Haydn, Mozart and, especially, Beethoven. The world has changed dramatically since their times, and as music and art reflect their own times quite nicely, our current music and art is reflecting something that Beethoven would not recognise. Our art has become awkward, disconnected and, sometimes, quite ugly, reflecting the situation around it. We often claim that we are living in the technological age, and yes, it is technology that enables much of our expression, whether in art, music or even the written word. However, there is no reason why the same technology cannot be used to create beautiful things and, sometimes, it is. One is always relieved when this happens as it shows that there remains a vein of gold among the dark, with individuals still capable of producing beautiful works. This quite often happens with traditional music. It is a shame that technology serves to smother much of this with the instantly constructed loops and samples which some consider to be music. Well, one could argue that there is a creative process involved, but it is not quite at the same level as Beethoven. These days, it is just too easy to create and distribute a composition, whether it has taken 5 days or 5 min to produce. Furthermore, in many cases, it requires no particular skill to do so. Those who do take particular care must try to find a suitable outlet for their efforts. There are various options, including the creation of 'made to order' CDs available from online outlets. And, of course, they may make their own CDs and distribute them as they wish throughout the community, but this is very different than having the backing of a large record company. Technology is sometimes a double-edged sword (Fig. 12.1).

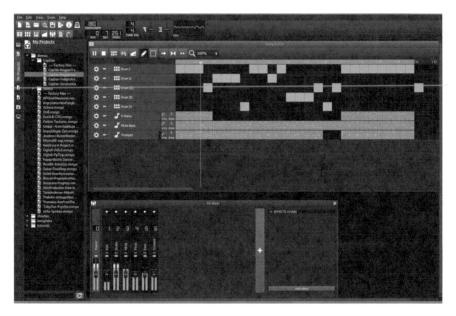

Fig. 12.1 An open source digital audio workstation

Chapter 13
Classical Music Is Effectively Broken by Technology

As previously mentioned, in the early days of tape recording, everything was fairly straightforward. Monophonic or stereophonic recordings were made, largely in concert halls or broadcast studios, with a minimum of equipment. Very often, it was simply a pair of coincident or near coincident cardioid microphones, positioned somewhere above the conductor's head and feeding directly into the tape recorder electronics, with the levels being adjusted on the tape recorder itself. Other configurations, sometimes with an array of three or four microphones, placed in a line in front of the orchestra, the signals running through a simple level and pan only mixer and then into the recorder, were also experimented with. The objective however was always to capture the sound of the orchestra, whether for direct broadcast, for recording purposes, or for both.

Indeed, this was the status quo for some time. Early 'high fidelity' enthusiasts who liked to build their own corner horn loudspeakers and spend many hours lovingly fitting their new turntable or valve amplifier into a cabinet that did not look too dreadful to the wives would enjoy the fabulous sound of an orchestra playing one of their favourite pieces, right there in their front room. It all seemed to fit together perfectly, and the enthusiasts would discuss their favourite recordings, almost entirely upon the merit of the performance. They did not worry too much about how the sound had been captured, as long as it sounded like the original performance, hence the popularity of the very term 'high fidelity' as in high fidelity to the original. If the Chicago Symphony Orchestra were playing in the Carnegie Hall, they should sound like the Chicago Symphony Orchestra playing in the Carnegie Hall. If the London Philharmonic Orchestra were playing in the Albert Hall, then that is what *they* should sound like. And, in those days in particular, enthusiasts understood the sound of different venues and thus knew what sort of sound to expect. The special treat was when an exceptional performance was captured. Those magical occasions when everything just seemed to go right. And this is where magazines such as *The Gramophone* played such an important part in providing consistently balanced reviews, upon which the enthusiast could rely.

J. Ashbourn, *Audio Technology, Music, and Media*,
https://doi.org/10.1007/978-3-030-62429-3_13

All of this, of course, is just how things should be. While, in the popular music field, the advent of multi-track tape recorders and large, sophisticated mixers and effects changed the sound and structure of the music itself, the effect on the world of classical music was, to start with at least, more subtle. Conductors and record producers found that they could 'play' with the sound after the event, especially if three of four run throughs had been recorded, as would happen with rehearsals and so on. A conductor might wish to replace the horn section in take 1 with that from take 2, and usually, the audio engineers could oblige. And then there was the question of level, especially with a concerto featuring a solo instrument. The temptation to lift the solo instrument in level, and maybe move it artificially in space, perhaps to be more central, was too strong for some producers. A good example of this is with Chopin's Piano Concerto No. 1 in E minor. The author searched for many years for a recording of this piece that seemed to strike the right balance between orchestra and piano, especially within those opening bars, and also maintain a seamless flow of time within that beautiful second movement. Even recordings by the big name pianists, hailed as 'milestones' or otherwise 'great recordings' seemed to get it wrong. The orchestra was too distant and pianists stumbled with the timing of that second movement, making what should have sounded beautiful, sound rather awkward. And then, by pure chance, a Hungarian radio recording, featuring an unnamed pianist was stumbled upon, and viola, there it all was. A perfect balance between orchestra and piano and a second movement flowed like a gentle mountain stream. Beautiful. Another fine recording of this piece may be found by Russian pianist Victoria Postnikova in an early Russian-made stereo recording where, once again, everything seemed nicely balanced.

So what had happened in the other so-called great recordings of this popular piece? Well, what had happened is that the balance between orchestra and piano had been interfered with, with the piano being artificially lifted. In addition, for some reason, the timing always seemed to go awry. Maybe this was the conductor's preference, who knows? But it was not *right*. The great pianist Solomon, who many would posit is the greatest recorded pianist of them all, would always take great pains to play what was written on the score. He believed, passionately, that it was the pianist's task to play what the composer intended, not his own 'interpretation' of it. The author has a particular dread of that word 'interpretation'. Beethoven, Mozart and others need no interpretation. No one is going to play it any better than they wrote it. And yet, in modern times, one hears soloists and conductors alike, proudly claiming that they have 'interpreted' a particular work for us.

The burgeoning technology within the control room made it even easier to 'interpret' things. Indeed, you could interpret until the cows came home, and many producers and conductors seemed to do just that, lifting a little here, cutting a little there, playing this section again only faster, and so on. Things were starting to become messy. Sooner or later, it was clear that this way of working would become the norm, especially with some of the big labels, and indeed, this was to be the case. Furthermore, some of them took it to extremes and produced some very clear sounding, but very artificial recordings. Of course, many people have never heard anything else and so believe that this is what classical music really sounds like. They will be in for a shock when they attend their first live concert, but, hopefully, a delightful one.

Fortunately, there are always exceptions, and some of the more specialist labels took pains not to fall into this trap. They were often the labels who would also commission and record the less mainstream works and, consequently, might not come to the general attention of the mainstream record buying public. The overall situation thus became a little confused and the continuum of the enjoyment of classical music performance according to how the music was originally written had been broken. Now, it was a free for all with people making records; however, they liked and performers often performing however they liked. We entered into the age of the 'popular star' classical performers and conductors, who focused on their appearance and on-stage antics rather more than being a servant of the composer. I can think of a few modern soloists who fall into this category. This is not to say that they are not entertaining, in their way, but they are not continuing the tradition of classical music and *that* is a big problem. The arts must surely be preserved as they were created. They are an important part of our history and our legacy. Beethoven did not intend that his piano concertos be 'murdered' by some flashy player who shows off in the easy sections and then fails to carry the more difficult parts. He wanted us to hear them as he had written them. That, after all, is how he is able to communicate with us over the centuries, and that 'communication' is priceless. If the technology further mangles his intentions, then what we are left with is a caricature of his works. This is a tragedy that we must surely strive to avoid. Fortunately, there are those that do, but it does mean that the record collector (via whatever medium) needs to be selective in their choice and to develop an understanding of who does what and who does it well. One gets a feel for this after a while. Let us hope that the pendulum starts to swing the other way and that more record labels produce true stereo recordings with the correct orchestral balance. To use an analogy with the art world, suppose we started to 're-paint' the paintings of Van Gogh, Renoir, Rembrandt, Michaelangelo, Botticelli and so on, using readily available image manipulation software. Certainly, we could create our modern 'interpretations' of the great works, but we would have destroyed the originals in the process. We should think about music in the same way and ensure that it is performed and recorded as the composer intended. A currently topical instance here is the tempi that Beethoven wrote on his manuscripts, sometimes retrospectively, after the metronome had been invented. Some claim that the tempo markings are too fast. But by whose judgement? Assuming that the metronomes of the age were reasonably accurate, then, surely we can take Beethoven's word for it. If one listens to Herbert von Karajan's 1962 recordings of the Beethoven symphonies, they are played at a faster tempo than the later recordings by the same conductor and many critics prefer the earlier examples as a more accurate representation of the works. If a composer provides explicit instructions in such matters then, surely, we should follow them. The same goes for recording and ensuring that we capture an accurate sound of the orchestra playing within a given environment.

Chapter 14
Digital Arrives, But Something Is Not Right

After years of becoming used to scratches and pops on shellac and vinyl records, almost out of the blue came digital recording and the compact disc, pioneered by Philips and Sony. It seemed too good to be true and was often marketed as 'perfect sound forever'. Naturally, there was a rush to build CD players and an equal rush to reissue record label back catalogue on CD for, very often, this is what happened, there were relatively few digital recordings being released in those early days. One reason for this is that people were trying several different formats and specifications with some quite bizarre suggestions cropping up. The author remembers some prototype recording set-ups, often using video tape and helical scanning, and some playback equipment that was the size of a small washing machine. Demonstrations were given within the industry which sounded fairly impressive in isolation, but there was no obviously clear direction. At Sony, Heitaro Nakajima had a passion and a vision for digital recording and was making progress even in the mid-1960s. Sony's early prototypes were also the size of washing machines, using multi-channel fixed head technology. It seemed that everyone was keen on bringing digital recording and playback to market, but how? The main issue was that of media and the fact that you needed, roughly, around one hundred times the space for digital audio than you did for analogue.

Fortunately, there was the parallel domestic market for video recorders, and this was blossoming with both Betamax and VHS. One of Sony's audio design team members, Akira Iga, realised that these readily available products already had a high enough bandwidth to record digital audio. Therefore, it was feasible to separate out the encoding and decoding into a separate box and then use this with an off the shelf video recorder. The result was the Sony PCM-1 which became available in late 1977. No more washing machines, the PCM-1 was small enough to be easily taken around, together with a Betamax video machine, to concert venues. Others were also making progress. JVC and Denon had been making digital recordings, and of course, Philips was heavily involved in understanding the commercial possibilities of digital.

J. Ashbourn, *Audio Technology, Music, and Media*,
https://doi.org/10.1007/978-3-030-62429-3_14

However, the PCM-1 was not universally praised, and many had issues with the early machines. This caused a review by the development team and some consequent modification. But Sony were also very lucky in that Herbert von Karajan, the famous conductor, was a friend of Sony chief, Morita and, at Morita's house, was shown the PCM-1 playing back a recording that the Sony team had made of a Karajan rehearsal. The latter, who was always interested in audio technology, was impressed and immediately realised the possibilities that lay ahead for digital recording and playback. Remember, that these were also the days of the video optical disc (sometimes called the laser disc), developed by Philips but also marketed by Sony. The video disc was quite large, about the same size as a long playing record, but you did not really need all this space for audio. Logical conclusions started to surface in both camps. Sony engineer Doi was spearheading the audio group and had the innovative idea of using error correction for the digital signals as, after all, this remained a relatively untried system. In 1977, various systems were demonstrated, mostly using the large laser disc. But Sony's system did not rely on video technology. It was a pure digital audio system. In June 1978, Sony engineer Ohga visited the Philips plant in Eindhoven, to find them well advanced in their own research for digital recording and playback. But Philips were working on a 9-cm disc. They wanted something that could be played in automobiles and had already calculated that, with their system, you could get 60 min of music on such a disc. It was felt that the two companies could usefully work together to perfect a small-sized digital disc that could replace the long playing record.

There had been a flurry of activity to, firstly, show that digital recording was a realistic proposition at all, and, secondly, to develop a playback product for the consumer market while simultaneously developing the art and science of digital recording. It all happened quite quickly, and many, with hindsight, believe that in this rush to get things to market, the wrong decisions were made, especially around sample rates and bit depth. However, by April 1980, standards had been established. Still there were issues as, tentatively, products started to appear at technology shows from around 1982 onwards and, very soon, would appear in high streets around the world. Perfect sound forever was not quite right, but it was certainly interesting and, just as certainly, caught the public imagination. Everyone wanted a compact disc player in their set-up.

But high-fidelity enthusiasts are a picky bunch, and quite soon, observations about actual sound quality started to surface. All were grateful for the absence of pops and clicks, but many noted that the sound had a 'hard' edge to it and that high frequencies in particular seemed somewhat ill-refined. If one, for example had a good quality analogue recording of the great violinist David Oistrakh playing something from the violin repertoire, one was enveloped by that beautiful, sweet tone that he coaxed from the instrument, as well, of course, as his peerless playing. But playing the same recording on early CD was not the same at all. The sweetness had gone. Replaced by a clean, but mechanical sounding presentation. Bear in mind that early CD players were still experimenting with beam technology, bit streams and other factors which would all come in to the final listening equation. In addition, as mentioned, many early CD releases were just duplicates of what had already been released on vinyl, together with any compromises that had been made with the original recordings, and so the potential of digital audio was not really being reached, while comparisons with analogue were all too easy to make.

It is a pity that a little more time was not taken in understanding digital audio and then a specification drawn up which would ensure its longevity as the hardware developed. The main bone of contention in the early days seemed to be the 44.1 kHz sampling frequency chosen for digital audio, at least, this was the popular scape-goat, together with the 16 bits of dynamic range with, in retrospect, was a little on the tight side for full-scale symphonic works. The 44.1 kHz sampling rate was set-tled on according to the Nyquist theorem which states, in simple terms, that when sampling and digitising a signal, the highest frequency which can be represented accurately is one half of the sampling rate. Therefore, 44.1 kHZ should give us up to 22.05 kHz in high frequencies which, surely, would be enough for any pro-gramme material, given the theoretical human hearing range of up to 20 kHz. A reasonable assumption. However, there were those who claimed that they could hear, or otherwise feel, musical harmonics above this frequency. In fact, if a spec-trograph is examined of a concert orchestra playing typical material, much of the energy being produced is well beneath these frequencies. The original specification for CD was considered quite reasonable by all those involved in setting the standard. But markets move quickly and other things began to occur.

Having effectively displaced the long-playing vinyl record with CD, Sony and Philips were thinking about a replacement for the humble compact cassette which had become very popular over the years, partly because of its universality and ready availability of both recorded and blank media. In addition, high-fidelity cassette decks had gone from strength to strength and, by now, were actually very good if good quality tape formations were used. And so, Philips launched the digital com-pact cassette (DCC) which was a well thought through system, providing back-wards compatibility with compact cassettes while also providing digital recording and playback using a data compression system. But Sony launched, almost simulta-neously, the MiniDisc which had nothing in common with cassettes and used a dif-ferent data compression system. So, the market was immediately launched into standards wars all over again, with some record labels supporting one system and others another. Most consumers kept their hands in their pockets and waited to see what would happen next (actually, the author quite liked MiniDisc for its physical robustness and often used it to record radio programmes). And then, to confuse things still further, in 1987 Sony introduced their own digital audio tape (DAT) for-mat which could sample at up to 48 kHz and which did not rely upon compression algorithms. DAT was intended as a consumer product, but the early recorders were quite expensive and the record industry was voicing concerns around consumers being able to make identical digital copies between DAT machines (but then, they could always copy compact cassettes, albeit with a slight quality loss). Consequently, DAT never really took off in the consumer market, but it was adopted in the broad-cast and professional markets for both its sound quality and convenient size—you could store a lot of DAT tapes on a small shelf. The author remembers once meeting with a BBC executive at Broadcasting House and being surprised to find his office liberally lined with DATs.

It occurs that there had been yet another possibility. In 1976 a consortium of Sony, Panasonic and Teac launched the Elcaset. This was a cassette tape cartridge, around twice the size of a compact cassette with quarter inch tape (twice the width of the compact cassette) running at three and three quarters of an inch per second (twice the speed). The resultant sound quality was very good indeed, especially if used with properly aligned noise reduction. The author remembers listening to a Sony Elcaset machine and being very impressed. This could well have been the tape format of the future and may have even formed a platform for digital audio, perhaps with helical scanning. Certainly, one could have built an easily transportable machine that would have been smaller and lighter than a Betamax video player and PCM-1 combination. Furthermore, the tape cartridges were rugged. However, there seemed to be little interest in the Elcaset and equally little marketing from the organisations involved. That was a shame as most consumers enjoy the tactile experience of placing a tape in a machine and adjusting the recording levels. If they are also given bias adjustment and the ability to align the tape properly with the machine, they have a wonderful device with which to experiment. Elcaset could probably have been adopted for several professional applications as well, but it was not to be. There was also the infamous eight-track cartridge, used mostly for automotive purposes, but that was a different kettle of fish altogether.

While all of these mixed views around standards were being aired, with format wars running along nicely, it started to occur to people that, once you had a digital signal, it did not really matter how you stored it and so, many of these discussions were quickly becoming irrelevant. This shall be considered later in this work.

But there was another issue with trying to raise the sampling rates of digital audio and that was simply that most of the analogue electronics around at the time were never designed to handle such high bandwidths. Transistor amplifiers of the time would not welcome high energy at high frequencies. Neither would the high-frequency drive units in most loudspeakers. Consequently, even if you had recorded super high frequencies onto digital media, it would almost certainly require high-frequency filtering before being played back on typical high-fidelity systems of the time. If memory serves, Harman Kardon were one of the first amplifier manufacturers to appreciate the need for wide bandwidth amplifiers that could be used with a variety of digital components, but most were not really thinking about this. Bear in mind that separate digital to analogue convertors had started to appear, some of them capable of going beyond the CD standard of 44.1 kHz. The world of digital sound was exciting. There was a fair amount of snake oil, but also some genuinely good ideas surfacing.

However, there remained opponents to digital sound, and record companies were not too keen on the idea of having to build CD manufacturing plants. Nevertheless, Sony, in particular, continued with an aggressive programme of constant PR, slowly getting recording artists on their side in the process. But how to get digital into existing analogue recording studios also presented a challenge as these big studios had already invested massively in high quality analogue technology. A breakthrough came when Sony introduced the PCM-1600 in 1978. This was based upon the U-Matic video platform and was a lot neater, and better, than previous incarnations. Suddenly, it looked reasonable for the recording industry to try their hands at digital recording, even if only for two track mastering purposes. Many of them did just this, combining the best of analogue and digital. Eventually, Sony would produce the PCM-3324 24-track digital recorder, and then, everyone was happy.

Well, not quite everyone. There remained the analogue diehards who simply refused to accept that digital sound could, actually, be made to be very good. Ironically, after the compact disc had almost finished off production of vinyl records, a new wave of interest saw the return of turntables (at higher prices) and all of the analogue paraphernalia such as specialist tone arms, pickup cartridges, phono pre-amplifiers, special connecting cables and more. And now, there are plenty of companies making vinyl records, some of them on heavier grade vinyl and some at ridiculously high prices. What had been underestimated here was the hobbyists' love of tinkering and custom building. There simply was not the same scope to do this with CD replay, other than use your own digital to analogue convertor (DAC) and play with connecting cables.

These days, a good recording, properly mastered to CD can sound superb when played back on good quality equipment. The original standard may have been a compromise, but it does work. On the recording side, audio engineers will typically record in high-resolution sound and then down-sample the resultant files for CD production. This is a technique which generally yields good results, although there are all sorts of views on down-sampling. Many engineers simply record in 24-bit and at 48 kHz sampling as this is something of a de facto standard with broadcast video. In reality, in most people's living rooms, the acoustics of the room itself, coupled to the quality of playback equipment being used, will make a bigger difference than worrying about sample rates or keeping the analogue/digital debate going. The truth is that digital sound and CD players have evolved over the years and are certainly capable of reproducing high-quality sound within a domestic environment, providing that the recording was undertaken properly in the first place. The author would like to posit that *this* is where we should be focusing our efforts at the present time. We may not have 'perfect sound forever' but we do have a very good, well-established format for reproducing music in the home. Digital works. What we need now is high-quality true stereophonic recordings (Fig. 14.1).

Fig. 14.1 CDs were
readily accepted by
consumers

Chapter 15
A-D and D-A Convertors and Compressors in the Digital Domain

One might posit that the most important point within the recording chain is the analogue to digital conversion, and the most important point within the playback chain is the digital to analogue conversion. If these functions are undertaken incorrectly, or with misaligned or with otherwise sub-standard equipment, then errors may well occur. The CD system has error checking built in, as do many digital recorders, but there are errors and errors. One does not wish to encode them into the signal in the first place.

At the analogue to digital conversion stage, the convertor will effectively slice the audio into discreet slices of time, giving each slice a particular value. If we assume a perfect analogue signal that is phase coherent and without distortion, then the audio will be properly represented by a numeric value within an available range. This range is the bit depth, such as 16 bits, 20 bits, 24 bits and so on. The greater the bit depth, the greater the dynamic range of audio that may be represented digitally. A perfect analogue to digital convertor would always output the same numeric value for a particular slice of audio at a given bit depth. However, as the great man himself (W. C. Fields) once said, there is nothing in this world that is perfect. So, a poor quality analogue to digital convertor may output a different numeric value for the same slice of music than a better quality device, even if they are using the same algorithms. Digital recording equipment often has integral analogue to digital convertors, and so the audio engineer is stuck with what he has within his equipment. However, he can give it a head start by ensuring that the power supply is clean (one may also power everything from batteries, which gives additional advantages in the field). Then, providing that no distortion is present within the input analogue signal, the A-D convertor can perform its function. Once it has done so, the resultant digital stream is all that the audio engineer has to work with, and so it had better be right. It goes without saying then that this analogue to digital conversion stage is critical to making good quality digital recordings, and this is why some A-D convertors cost more than your car.

J. Ashbourn, *Audio Technology, Music, and Media*,
https://doi.org/10.1007/978-3-030-62429-3_15

The reverse process, to convert the series of digital values back into an analogue sound wave that we recognise as the original recording, is equally important. There are companies who do nothing but manufacture these digital to analogue conversion chips, and the competition between them is strong. Device manufacturers tend to have favourite DAC chips, but sometimes alternate between models. Price is also a deciding factor of course when designing a digital device that will use one of these DAC chips. Such a device would typically be a digital player of some kind, but there are many standalone digital to analogue convertors for enthusiasts to place in line with their CD transports, or other digital sources, and amplifiers. In fact, many amplifiers also have in-built DAC capabilities. Similarly, portable digital audio players also use the same chips from the major suppliers. Consequently, DAC chips are well developed and mostly provide a high performance and are capable of decoding high-resolution signals, often up to 192 kHz as well as DSD streams (as were devised for the ill-fated super audio compact disc). An interesting factor here is that the manufacturers of these DAC chips tend to market them on sound quality. Well, if one sounds different to another, and they do, then it indicates that none of them are decoding the signal properly but that some seem to make a better job of it than others. The end result will, naturally, also depend upon the specific implementation of the DAC within the overall design of the device in question.

It is clear then that A-D and D-A convertors lie at the very heart of digital sound. They both need to do their job properly if we are to enjoy good quality. At present, they are doing the job, but there are variations between them, and this makes digital recording, in particular, very interesting. Audio engineers will have listened critically to recording equipment and will have their own preferences, of course, and certainly there is some very interesting equipment available to them.

The situation within recording studios, especially those focusing upon popular music, is slightly different as they seem to place a great importance upon the software they are using, its facilities and how easy it is to 'plug in' digital effects or sampled instruments. When recording live instruments, there is a tendency to place what is known as a 'digital interface' between the instruments and the computer. This interface may be an integral part of the mixer being used or may be a standalone item connected to the mixer and to the computer via a USB port. In any event, this is where the analogue to digital conversion is taking place. Some of these devices are inexpensive and seem to be sold simply on their bit depth and sample rate values. There will no doubt be significant variations in audio quality between them.

There is another factor that affects digital recording which is worth mentioning. On conventional analogue mixing desks, it is a common practice, especially in the popular music field, to add compressors and limiters into the signal path in order to control unexpected peaks in the music. In the popular music field, these devices are used somewhat excessively, in order to make the music sound louder. Someone recording a live event, such as a classical concert, might like to switch in a compressor or limiter for its correct purpose, to control unexpected peak levels and so avoid distortion. A compressor, as the name suggests, gradually compresses the audio once it has passed a set threshold, but it will not stop it from going into distortion if the signal is strong enough. A limiter is a hard-edged filter which simply stops the signal from going past a predefined point. The problem is where these devices are functioning. If they are in the analogue domain, then they will perform their function before the signal is subject to analogue to digital conversion. However, in many affordable digital recorders, they operate in the digital domain, after the signal has been converted. This is no good at all because it means that the original distortion has been encoded into the digital signal and cannot be removed. The best option in such cases is simply to turn all dynamic processing off and record at 24-bit depth, thus allowing plenty of headroom for the incoming signal. There is a similar issue with software-based compressors and limiters. If the signal they are being used on has already been digitally encoded, then any distortion within the analogue signal will have been encoded along with it. However, in the software world, such devices are often used purely as an effect. There are various ways in which components may be tied together, whether in hardware or software, and this needs to be taken into careful consideration when recording.

In conclusion, A-D and D-A convertors may sound different from each other. Critical listening is the way to decide what is best for a given application. Once the signal has been converted to digital, then that is that. Any work you perform on the sound now will be undertaken within the digital domain and, of course, can affect the sound quality as you will be altering the original digital signal. When it is finally converted back to analogue, it will surely be different. Whether one is concerned with this reality may depend upon what one is trying to achieve. If 'high fidelity' is the goal, then the less processing that takes place, the better. Getting the sound right at source is far preferable than trying to alter it after the event. The digital process makes this maxim truer than ever (Fig. 15.1).

Fig. 15.1 A separate digital to analogue convertor

Chapter 16
High-Resolution Digital Recording and Re-Sampling

The dissent voiced about the sound of early digital recording and the specification of the compact disc never quite went away. Those who thought that a different sampling frequency might have usefully been used, and in conjunction with a greater bit depth, continued to experiment with what became known as high-resolution digital audio. They built special analogue to digital convertors and were soon able to make high-resolution recordings, of one sort or another. But to what end? The CD had quickly become established, and most people were happy enough with the sound they were hearing. Even the DCC and MiniDisc formats, which both used data compression techniques, with an associated drop in absolute sound quality, were finding favour among enthusiasts. Furthermore, realising that you could store digital sound on anything that could hold data, the MP3 compressed format, developed by the Moving Picture Experts Group as a way to store images and sound, was also finding favour. So why all the fuss?

It is often the case that, once an idea takes route, it does not go away, and this was very much the situation with regard to high-resolution digital audio. DAT tape had already increased sampling up to 48 kHz and people generally liked the sound of a clean DAT recording. So, maybe it was time to go further. The problem was, of course, that the CD standard had already become established, and so, even if one were to make high-resolution recordings, how would consumers play them back? Especially given that most had high-fidelity equipment designed for analogue sources. There was also the question of dynamic range and bit depth. Sony addressed this with something called Super Bit Mapping (SBM) which sought to squeeze a 20-bit recording into the 16-bit space of CDs. It seemed to work well enough, but I doubt that the average consumer could tell any difference.

While all of these discussions and experiments were taking place, those with good research and development facilities were working behind the scenes. Eventually, in March 1999, Philips and Sony agreed a standard for the Super Audio Compact Disc, which also featured strong copy protection, thus appealing to the record companies. The SACD disc also allowed for single-layer, dual-layer or

J. Ashbourn, *Audio Technology, Music, and Media*,
https://doi.org/10.1007/978-3-030-62429-3_16

hybrid formats, allowing the disc to contain both the high-quality SACD track and a compatible CD track, providing the universality that consumers liked. It all should have been fine, with device manufacturers slowly moving to SACD players and the record companies releasing plentiful output of SACD discs. But it simply did not happen. The early SACD discs were expensive, sometimes almost twice the cost of their CD counterparts and, in addition, there were few players available, and they too were expensive. Consumers kept their hands in their pockets again and people slowly lost interest in SACD, even though it was truly a very good format. The manufacturers and record companies had been greedy and effectively put up a barrier to entering the world of SACD. The author has long felt that this was a great shame as SACD really did offer interesting possibilities, including for multi-channel sound. There was also an attempt to launch DVD audio, but this met with similar disinterest.

In the meantime, those who were interested in high resolution kept going. They realised that separate DACs and various methods of data storage would make high resolution a reality for those who wanted it. Manufacturers started to see this as well and provide hard disc storage systems which could be incorporated into existing high-fidelity systems. Of course, the entire system had to be capable of handling high resolution, and so we started to see the introduction of loudspeakers with the ability to produce output at 30 kHz and above (which is not necessarily very useful) and wide bandwidth amplifiers 'designed for digital', or so they claimed. At the other end of the market were impossibly cheap 'MP3 players' which featured integral memory and the ability to play both compressed and uncompressed music. At the lower end of the market, these were very basic devices, but there was another market for very high-quality portable digital audio players that could be used with equally high-quality in-ear monitors. These devices could play a wide variety of formats in resolutions that crept upwards from 96 kHz to 192 kHz as well as the ability to play DSD files from SACD. It was as though the consumer market had neatly side-stepped the ideas of the big manufacturers and created their own 'high resolution' market place. They could 'rip' files from their existing CDs to their portable devices and also download high-resolution audio files from specialist sites on the Internet. Device manufacturers made 'streamers' with which such files could be fed into existing audio systems, and a whole new market emerged. But was it really the high-resolution audio that was originally sought?

There are two main factors with high-resolution audio. The first is the increase in high-frequency response. At 44.1 kHz sampling rate, the high-frequency response extends, theoretically, to 22.05 kHz, which should be more than enough for most programme material. At 48 kHz sampling rate, it extends to 24 kHz, a useful extension. At 96 kHz, the high frequencies extend to 48 kHz which is beyond useful as you do not really want high energy going through your system at this frequency. At 192 kHz sampling rate, the theoretical high-frequency response extends to 96 kHz which is absurd. In practice, low-pass filters are usually used in order to remove energy at these frequencies. The slope of these filters has a direct affect upon the sound, and modern DAC manufacturers often offer a selection of filters, from slow to fast, from which the user may choose. Slow filters often give a more realistic presentation while fast filters sometimes offer a better theoretical performance.

However, extended high frequencies are not the primary benefit of high-resolution audio. The real benefits lay in the increased granularity of sampling, especially within the middle and high frequencies, and the additional smoothness and detail that this provides within the signal. At a sampling frequency of 96 kHz, we are cramming twice as much information per second of audio as we are at 48 kHz. At 192 kHz, twice as much again. This moves us much closer to a true analogue signal. If we have a device upon which we can play these files, the keen ear will readily detect an improvement in sound quality. Many will hold, however, that there is little point going beyond 96 kHz, and of course, with every increase in sampling frequency comes an attendant increase in file size. Beethoven's ninth symphony recorded at 192 kHz would create a truly massive file, and while data storage is relatively inexpensive, moving these files around can take a long time. There is also the 88.2 kHz sampling rate, which makes for a nice clean down-sampling to 44.1 kHz for CD.

There are two schools of thought when it comes to recording specifically for CD. Some maintain that as any manipulation of the digital file, such as down-sampling introduces some form of distortion, it is surely better to simply record at a sampling rate of 44.1 kHz to start with and, therefore, not need to manipulate the file. Others hold that the increased quality of a 96 kHz or 192 kHz recording remains detectable, even when the file is down-sampled back to 44.1 kHz with 16-bit depth. The author has also experimented with this and feels that it *is* worth recording at high resolution and subsequently down-sampling. The resultant files, if handled carefully, do seem to retain some of that deliciously smooth mid-range and increased 'air' in the higher frequencies. Of course, one always has to handle digital files with care.

There is something of a question mark over some files which are tagged as 'high resolution' and which may be downloaded from some of the sites on the Internet. When these files are purported to be of popular albums which were released in the 1960s and 1970s, or even later, one wonders what these really are. Are they really high-resolution files produced from the original master tapes, many of which might not even exist now? Or are they simply up-sampled files from existing CDs? The user must exercise their own judgement in this respect. Certainly, some have been accused of the latter practice, in which case, the files may be in a high-resolution format, but the quality of sound will be no better than the source. If they are new recordings which have been recorded in different formats at source, then that is another matter and one should easily be able to distinguish between them. However, the consumer must choose accordingly.

Thinking of where high-resolution recording and playback might go in the future, it is clear that we have in place capabilities for both recording and playback in high resolution. This may also become increasingly important in the video world. The potential for moving forward may lay, in some respects, with down-sampling techniques. Consumers are still purchasing CDs, even though downloading music from the Internet has become popular. Consequently, there may still be scope to improve the realised quality of CD which, by the way, can be very good indeed. A well-recorded piece, properly mastered for CD can sound fabulous on the right equipment, and many audiophiles have invested in high-quality CD playback, ensuring that this remains a viable market. Improving down-sampling techniques may squeeze a little more quality from CDs. In addition, those who do like to download music from the Internet do not necessarily do so in high resolution. In fact, the compressed MP3 format still accounts for much of downloaded music. Others may simply download at 16 bit, 44.1 kHz and choose whether to burn their own CDs from the downloaded material. And some, of course, will download in high resolution. Whether they really get the benefits of high resolution may be questionable, but there it is.

Chapter 17
Lossless Compression

Another, quite useful development in the field of digital audio has been the provision of various lossless compression formats. Such formats are also used for images and simply for data compression, such as with the ubiquitous ZIP format. For audio purposes such formats include WMA Lossless, AAC, FLAC, APE and others. They all claim to be lossless, and most of them offer variable levels of compression. And so, the potential is that you may encode all of your CD files into a much smaller file size for storage, either on a portable digital audio player or simply for archiving purposes. Furthermore, some of them, such as the APE format from Monkey's Audio, claim to be able to perfectly recreate the original file from the compressed file should you ever need to do so. For example, if you damage or lose a valued and hard to replace CD, you should be able to regenerate the original files and burn a new CD for yourself. It all sounds very exciting and very useful, and indeed, these formats are very useful.

The underlying theory is to look for correlation between channels and to remove data redundancy. Often this is achieved by converting the stereo channels into mid-side channels and then using complex prediction algorithms to identify and remove redundancy. The variable levels of compression allow a balance to be achieved between real-time data processing and file size. However, the file size parameter is really becoming academic with today's data storage techniques, and most computers will easily cope with the data processing. However, on replay, portable digital audio players in particular must have an efficient data processing capability in order to play the most compressed files cleanly. In practice, this is rarely a problem.

There are many computer applications available with which to 'rip' the music files from CDs (or other sources) and render them in a lossless audio format. Some work more efficiently than others and several are freely available, including some of the better examples. Perhaps the most popular of the formats is the open source FLAC format, which has attracted the efforts of a large group of developers. It certainly works well, and well-rendered FLAC files sound very good indeed. Another popular format is the APE format from Monkey's Audio. Almost any portable

J. Ashbourn, *Audio Technology, Music, and Media*,
https://doi.org/10.1007/978-3-030-62429-3_17

digital audio player will play both of these formats well, as will several high fidelity storage and streaming systems.

And so, we have a good selection of lossless formats from which to choose which, in theory, should preserve the original sound in its full quality. We also have lossy formats such as MP3 which continue to be very popular, with many people listening to audio only via MP3 files played upon their smartphones. This, of course, cannot be considered as high-fidelity sound but, for many people, it is all they want. Getting back to the lossless formats, which may be considered as high-fidelity sound, there are some interesting points to consider.

Let us imagine that we have gone to great lengths to make a very high-quality recording at 24-bit depth and 96 kHz sampling rate. Our resulting digital file will be of size 'x'. Now, let us use one of these lossless formats to encode our file. The amount of redundant data which is discarded we shall call 'y'. Our file size is now '$x-y$'. We have lost 'y' amount of information. The question is, was that information *really* redundant or was it just *supposed* to be redundant by the lossless algorithm? If it really *was* redundant, how and why was it recorded in the first place? Mathematicians will take great delight in proving to you, beyond any doubt, that data compression algorithms work perfectly and may always recreate the original file, exactly as it should be. It is a strong argument, especially when working with conventional data and you can easily demonstrate it for yourself. But here is the really interesting point. If all of these systems are working correctly, then, within a blind listening test, you should not be able to tell any difference between them. Of course not, they are *lossless* after all. They should therefore all sound identical.

Well, the author has experimented quite extensively with all of the popular lossless algorithms and has discovered two important realities. Firstly, using the same algorithm at its fastest (least compression) and slowest (most compression) settings yields a detectable difference in sound. This is not really surprising as the fastest setting is creating a larger file containing more information. Exactly what this information is, is open to question. Secondly, encoding the same piece of music from the same source, but using the different algorithms reveals that the resulting tracks do indeed sound slightly different. So, if they sound different from each other, how can they be lossless? Surely, they should all sound identical? But they do not. You may like to try this experiment for yourself. I repeated the experiment using several different digital audio players, in order to eliminate any processing efficiency issues, and obtained the same result every time.

However, we are really splitting hairs here as the leading lossless algorithms do produce digital files which sound very good indeed and require little storage space. It is simply interesting that our assumptions about these things are sometimes usefully tested. The author has previously written a very strong data encryption programme which is designed in a way as to produce different cipher text every time it is run on the same plain text. But it always decodes perfectly. Of course, there is a good reason for this too. The digital world is full of surprises.

However, there may be much more that we can do with audio (and image) compression codecs in the future. We are now back in the world of surround sound and immersive audio and have some interesting ways of capturing, storing and replaying these files. It may be that lossless algorithms continue to be improved and become used across a broader range of applications. This would seem like a logical development. They have been proven at the relatively simple level of compressing an audio file down for easier storage. Perhaps there is still a long way to go. The suppliers of these lossless compression codecs may like to start experimenting with carefully controlled listening tests in order to arrive at a better perceived sound quality, especially that at the higher frequencies. No lossless format is perfect at present. These are interesting times for both the recording and reproduction of audio (Fig. 17.1).

Fig. 17.1 A digital audio player

Chapter 18
The Revolution in Playback Technology

The round black disc, first shellac, then vinyl, has been with us for a very long time, and there are entire generations who knew no other way of reproducing music in the home. The introduction of high-fidelity stereo, especially, brought a satisfying level of music reproduction into countless living rooms around the world. And the vinyl disc remains with us today, together with fancy, audiophile turntables, pickup arms and cartridges. There can be few people in the world who have not seen or heard music being reproduced from discs of one sort or another. However, as we have seen, various other formats have cropped up and, some at least, look like they will be around for a while yet.

The compact cassette, from Philips, was perhaps the first revolutionary product. Tape recorders already existed of course, but they required reels of tape which had to be carefully threaded through the guides, over the head block and through the capstan and pinch-wheel which controlled the record or playback speed. There were portable tape recorders too. Miniatures of their bigger brothers with five and a quarter inch or even 3-in. reels. Naturally, you did not get much recording time with these machines, and this was a problem for professionals, especially those gathering news. In addition, the quality was somewhat variable and you had to pay to get a good machine. The humble little cassette, with its one eighth inch tape and one and seven eights inch fixed recording speed, was intended really as a cross between a dictating machine and a portable hobbyists tape recorder, albeit of rather basic quality. Philips probably would have been happy to leave it at that, but others had other ideas for the fabulous little tapes which were so easy to handle. And so, portable machines started getting better and better (the author has two Panasonic cassette recorders which are 40 years old and still working fine). They were soon integrated with radios and, although the compact cassette originally had just two mono tracks, enabling either side to be recorded or played, why not fit in four tracks and make it stereo? And, while we are about it, let us put in some decent electronics and sell it as a hi-fi component? The idea seemed mad as the cassette was never meant for such things, but record companies were releasing pre-recorded material on cassettes and

J. Ashbourn, *Audio Technology, Music, and Media*,
https://doi.org/10.1007/978-3-030-62429-3_18

the public were lapping it up. And so, component cassette decks started to appear, including from Philips. The problem with these first crop of cassette decks was, naturally at that tape speed, hiss. Philips came up with their dynamic noise reduction (DNR) which worked tolerably well and was compatible with everything. And then Dolby B noise reduction was licensed to the big manufacturers and, all of a sudden, cassette decks did not sound like such a bad idea. Akai were among the first to get it right, but others soon followed and, as competition was strong, improvements in sound quality were rapidly made. The record companies started to release Dolby B-encoded cassettes and everyone was happy. Cassette decks sold like hot cakes.

When the compact disc came along, it was sufficiently different to capture the imagination of the record buying public. It looked different and it sounded different, but early discs were a little expensive compared to their vinyl counterparts and CD got off to quite a slow start, especially with questions being asked about digital sound quality. There were not that many in the record stores to start with either. But the CD was cute, and slowly, audiophiles started to by component CD players and, naturally, as more entered the market, the quality improved. Then there was a veritable boom as everybody wanted everything on CD. Record companies had never had it so good. They simply reissued everything on CD, while slowly getting to grips with recording in digital. There were fortunes to be made here, and the point was not lost. Consequently, the CD was also revolutionary, partly in itself and, partly, because it created a new market for standalone DACs and other accessories. There were even CD juke boxes, which were popular in some markets. Now, not only are there interesting new releases on CD, but there is a thriving used CD market in many countries, and although critics have been heralding the demise of the CD for ages now, it refuses to lie down and remains with us.

And so, the compact cassette and the compact disc were both revolutionary products which changed the way we listened to music. The digital compact cassette, the MiniDisc, the DAT cassette, the Elcaset, the Super Audio Compact Disc and DVD Audio have simply raised a few ripples in the marketplace but have not really taken root, even though the technology itself was revolutionary enough. Why was this? It is probably all to do with marketing and market placement. Not enough attention was being paid to what the buying public actually wanted. They like simplicity and things that work at affordable prices. They do not want expensive things which are complex and seem to duplicate what they already have.

These marketplace revolutions were based upon products and technology. But there was another revolution lurking that was not really based upon products at all, but the fact that you could quite easily store digital music on virtually any popular storage media, from computer hard discs to compact flash and SD cards. Furthermore, very low-cost products could be made to exploit this reality, and everyone was talking about compression algorithms. So, why couldn't I store music on my phone? You could. Why couldn't I connect a portable digital player into my hi-fi system? You could. Why couldn't I simply listen with lightweight headphones while on the move? You could. In fact, almost anything you could think of had somebody making relatively low-cost products to meet the requirement. And then, the digital revolution changed up a gear or two and exotic, very expensive components started to appear, and people bought them. Now we had products ranging from a few dollars to tens of thousands of dollars, all doing essentially the same thing, playing digital music. And everyone seemed to know all about it with very little marketing required. This was truly revolutionary. All we need to do now is raise the quality at the bottom level and encourage proper stereo recording for both live concerts and studio projects. It is a tragedy that, within this roller coaster digital ride, the art and science of true stereo recording has taken a back seat to multi-track, multiple microphone techniques which do not provide what the composers intended. In the popular music field, other than vocals, much of the music is not even coming from proper instruments but is produced from synthesisers wholly within the digital domain. In the classical and traditional music world, we need to cherish our wonderful legacy of music and ensure that it is reproduced as it was intended. In the popular music field, well, people will experiment, and occasionally, some nice things might break through. It is ironic that, currently, some of the best-selling popular music seems to be being made by those of the author's generation. You can spot us by our white hair and tottering footsteps. But there will be more revolutions coming. Hopefully, among these will be a revolution in true, realised quality at affordable costs and a return to true stereo or other ways of reproducing performances as they are actually heard in situ. There may be further improvements in microphone design. New materials and surface mount technology may play a part here. Similarly, in the construction of microphone cables, there is potential for improvement. And, of course, continued development of ADC and DAC components.

Chapter 19
The Social Revolution in Consumed Music

The previous chapter considered technical revolutions in the way in which we consume (and make) music. But there is another revolution which is just as real and, in some cases, quite worrying. And that is a social revolution which is almost global in scale. It has many facets.

If we return to the 1950s, recording stars, in America, were often film stars, crooners like Dean Martin, Frank Sinatra, Doris Day and so on. In England, they tended to be music stars, people like Kathy Kirby, Vera Lynn, Jimmy Young, Alma Cogan, Susan Maughan and the hard to place but unique voice of the incomparable Kathleen Ferrier. However, in England in particular, these were singing stars who *acted* like stars. They were impeccable in their appearance, in their manners and the professionalism with which they approached their careers. Members of the public would have their favourites, of course, but there was not the sort of senseless hero worship which we see today. People viewed their favourite singers in a different light. Kathleen Ferrier was considered the second most popular person in England, after Queen Elizabeth. People did not mob her, but they *loved* her. It is hard to find anyone who actually met Kathleen Ferrier who was not instantly smitten with her. The reason, apart from that most beautiful of voices, is that she was a sincere, modest, charming individual who was also intelligent and professional when it came to her performances. Stars were stars in those days, carefully managed by their record companies or PR agencies. They always appeared well dressed and fully prepared for either interviews or performances. It was all nicely organised and exactly what people expected.

However, in America, the home of jazz and excesses, there were strange stirrings. A new, unexpected style of singing and performing, loosely called 'rock and roll' suddenly took off, with Elvis Presley, hotly pursued by Jerry Lee Lewis, Little Richard, Bill Haley and a brace of others, all churning out rock and roll and giving 'over the top' performances which youngsters immediately warmed to, especially young ladies who would scream and jump at such performances. It has always been a sense of wonder to the author as to why the females of the species behave like this.

© The Author(s), under exclusive license to Springer Nature Switzerland AG 2021 81
J. Ashbourn, *Audio Technology, Music, and Media*,
https://doi.org/10.1007/978-3-030-62429-3_19

Perhaps it echoes the distant past in some way, who knows? Anyway, rock and roll swept through America like wildfire. It was not to everybody's taste, but it captured the moment with young audiences. It also made some of the performers very wealthy although, this often proved to be a double-edged sword, due to their inexperience of managing such things. There are countless stories of alcohol and drug abuse, and of course, this was commonplace in the jazz scene where, since the 1940s, it had almost been expected that performers would succumb to such temptations, and sadly, many of them paid the price with shortened life spans and misery along the way.

In England, there had been some musical quirks such as skiffle (good old Lonnie Donegan), variations on folk and one or two bands (like John Mayall's Bluesbreakers) tinkering with the blues. And British audiences readily bought the American records. The record industry decided to support a few English youngsters who were emulating, but only to a degree, their American counterparts. Acts like Cliff Richard, Billy Fury and a few others were properly supported with good quality production, often sporting orchestral backings. And we still had the likes of Kathy Kirby, Susan Maughan and others. There were also the occasional novelty songs such as 'Right Said Fred' from Bernard Cribbins, and 'I'm Walking Backwards for Christmas' by the Goons. It was a fascinating time in England, with TV just coming along, and some individuals who had been turned down by BBC Radio found a welcome at the fledgling BBC TV. One such name is David Attenborough and another was a young lady with the most perfect, cut glass English accent that you could imagine. Her name was Muriel Young, and they gave her a children's programme on early BBC TV (named 'Tuesday Rendezvous'). The reason I mention Muriel is because her programme was chosen, tentatively, to show a new English band who had recently been signed to EMI. They wore identical, trendy suites, with ties and were characterised by their mop like hairstyles. They called themselves The Beatles. All of a sudden, England had its own hysteria stirring act, and the Beatles could do no wrong. But there were others too, equally well-groomed and writing catchy songs. These included The Searchers, The Hollies, Cliff Richard and others. It was all very civilised—for a while.

But then came a new, explosive wave of British popular music. It seemed that anyone could buy a guitar, learn a few chords and try their luck. And hundreds of them did. Some fell by the wayside and some never took it too seriously anyway. But some were aggressive and fiercely competitive. Most managed to get deals with the major record labels, and those that did not were often welcomed by new labels, also managed by the younger generation. The problem was that they had lost the well-groomed and carefully stage managed performance of the 'official' popular stars and everything became a free for all, with excesses of all kinds being practised. Furthermore, while there were some genuinely catchy tunes emerging (Ray Davies from The Kinks became a renowned songwriter), there was also a great deal of self-indulgent rubbish. But the volcano never seemed to stop erupting, and a huge number of English bands were formed, churning out an endless stream of recorded music. There were open air pop concerts and all manner of wild parties and, of course, alcohol and drug abuse soon left its mark. But it was also an exciting and very productive time, not just for music but for all industries affecting the younger generation. This was the 'swinging sixties' at full bore, and everyone felt its effect. In the music industry, not only were there the young working class to riches performers, but the same was happening with producers and those producing themselves. In America, there was Phil Spector, Frank Zappa and others experimenting with sound, using compressors and other effects to create an identifiable sound while, within bands there were those like Brian Wilson of the Beach Boys, obtaining a sound of a different sort. In England, George Martin was helping the Beatles along at EMI.

But also in England, in a flat above a shoe shop in Holloway Road, London, was an eccentric tinkerer. He lived and worked among a shambles of tape recorders, pieces of mixers and a range of effects boxes which he had designed and built himself. Amid shouts from downstairs to keep the noise down, he played with his equipment and managed to produce some unique sounding songs. Some considered him a flawed genius and his life, together with that of his landlady, eventually ended abruptly and tragically. But, in the meantime, after dismissing an early demonstration tape from the Beatles, he had a No. 1 hit with the unlikely film star John Leyton (goodness knows how the two of them ever got together), followed by the international hit 'Telstar' and several other successes. His name was Joe Meek.

Things were never the same after the 60s and there were several distinct 'movements' within the popular music arena, including the so-called punk period which seemed to be punctuated by anger, bad language and bad behaviour on and off stage. Some consider this a creative period, but what was it creating? Echoes of this remain today with some music being nothing but a repetitive rhythm with angry shouting over the top. Of course, there are nicer strands as well, but, overall, popular music seems to have become angry and ugly and is also starting to be used politically, which is certainly a shame.

The reason why such, probably minority, strands prevail and are blasted at us from every direction is because music has become instant. A band could record or otherwise make a digital 'record' at 10 a.m., place it on several channels on the Internet at 11 a.m. and, by 12.30 p.m., the likelihood is that it will already have been heard by many thousands of people. By the same time the next day, this figure may be in the millions. And so, those individuals (usually the most outrageous) who wish to exploit this will do so ruthlessly. Others may be a little more concerned about security on the Internet and other areas of social responsibility. And, of course, music may be placed on 'drop' sites and widely distributed that way. In addition, staying in the popular music field for a while, some of these acts are signed to the big labels and their music distributed via more conventional methods. So, we have a free for all whereby, it seems, that almost anything may be considered as music, even if it has no structure, no melody, no harmonies and almost unintelligible lyrics. Such is the power of technology. But more serious musicians and songwriters have to exist also within this technology maze, and some will no doubt emerge through it. What is missing is the valuable artist and repertoire screening that the big record companies used to perform, together with the high production standards upon which they insisted.

In the classical music field, the music has not changed, but the way it is sometimes played has. We are in the era of the superstar classical soloist, and, sometimes, such individuals seem to want to 'show off' a little and show us how fast they can play a particular piece, even if they are not playing it correctly. Some might call this 'showmanship', but there is really no need for it in classical music. There are some individuals who also insist on dressing oddly, rather than conforming with tradition, and this is not limited to the players, but conductors also. One is tempted to suggest that the best way of drawing favourable attention to themselves is to play what is on the score.

Classical music concerts and opera are also struggling, in some countries at least, with the cost of performances, and this is reflected in ticket prices which, some feel, are too high. However, staging a full programme classical concert is no mean feat and requires a good deal of work and expense before the event. And so, there is a culture change here which is in danger of pushing live classical concerts out of reach of the masses. It is a difficult problem, and one may only hope that some sort of equilibrium is to be found. Regional and amateur orchestras may be more flexible in this respect, and perhaps, we shall see some good things happening in this area. However, the social revolution in how music is now produced and distributed is certainly having an impact. A massive impact in the case of popular music and a lesser one in the case of classical music, except that, even here, the possibility of downloading individual movements or excerpts and then streaming the files onto either fixed or portable devices might well impact people's understanding of classical music. Some may not have the attention span to critically listen to entire pieces or the interest to understand the history of the piece being played. This is noticeable on some of the DAB broadcast stations where presenters sometimes 'interpret' short passages of music, giving misleading information about its history or the composer concerned. This is a tragedy as a good, in-depth understanding of classical music is one of the greatest attributes that an individual may acquire in their lifetime. Helping them to do so should be our aim, as it is also one of the most profound of pleasures in life. It goes without saying that the performance of this music is also critical and some would hold that orchestras and musicians in general were better in the 50s and early 60s than they are now. It is a thorny question, but you can understand why it is raised. There are some excellent individual musicians around today, but the ability to play well in ensembles is another skill. This brings the matter of training into question. Everyone wants to be a star, but the brightest stars are those who serve the composer and the score.

Chapter 20
The Change in Musicians

Have musicians really changed? Surely, playing Mozart now should be the same as playing Mozart 50 or 100 years ago? After all, the notes have not changed. Musical instruments may have changed slightly, but a violin today surely sounds, at least very similar, to a violin then? All of these arguments are entirely reasonable and, if we were to posit that there has been a change, then what do we have as evidence? Well, of course, we really only have good evidence from the time that such music has been recorded, plus some more tenuous evidence in the form of commentary of performances long ago. Think of the violin virtuoso Niccolo Paganini for example. He was apparently an extraordinary performer, and some maintained that he had sold his soul to the Devil in exchange for his musical talents. He was, by all accounts, a superb violinist, and it is a pity we cannot hear him today. Beethoven was, especially in his younger years, a superb pianist and had no equal in his powers of improvisation and superb control of the instrument. But we cannot hear him either.

What we can hear are very early recordings from the 1930s and 1940s, and some wonderful performances from the former communist countries state radio broadcasts as well as early specific recordings made in the 1950s and 1960s. Also, early recordings from around the world from the same period. If one listens critically to these recordings, it quickly becomes apparent that both the playing and the overall sound are somewhat different from the sound of modern orchestras. This is not just a question of available technology. They are playing differently. The tone and overall musicality of solo instruments within concertos seem much sweeter and more communicative. The orchestras are coherent and *sound* like orchestras and provide a good sense of 'being there'. Furthermore, the balance between orchestra and solo instruments is more realistic within concertos, and full symphonic works convey a realistic sense of dynamics. Live Opera recordings from this period are also very special and convey all the spatial information of the venue, including movement of the leading players, together with realistic dynamics. The lead roles being sung with the passion that they deserve by the leading artists of the time.

© The Author(s), under exclusive license to Springer Nature Switzerland AG 2021   85
J. Ashbourn, *Audio Technology, Music, and Media*,
https://doi.org/10.1007/978-3-030-62429-3_20

Things are simply not the same today, so what has changed? There are perhaps two areas which can be seen to have changed. Firstly, the fine dissection of time and dynamics by musicians of this era and, secondly, the attitude of the performers themselves. Let us examine these factors a little more closely. A really good musician, in any field of music, has the ability to dissect time with minute precision, playing either slightly behind or slightly in front of the note at certain points, even sometimes within the same musical phrase, in order to provide the 'movement' and 'flow' that the piece requires. For example, if one were playing one of Chopin's wonderful Nocturnes, a really good pianist, who would know Chopin's own history and personality, would try to really get inside the music and play for you as Chopin would do himself. These are romantic pieces that require an appropriate level of skill and understanding in order to really make the piano sing and communicate Chopin's message.

In musical notation, there is such a thing as a demisemihemidemisemiquaver which is a 256th note, but composers rarely go down to this sort of granularity. In practice, a semiquaver (16th note) or demisemiquaver (32nd note) will serve the purpose within most scores. But experienced musicians will understand that it is not just the note value but its *precise* location that gives the music its sense of movement and flow, and that, sometimes, such things are implied. Paganini would have understood this. It is the same with dynamics. Music notation gives as piano (soft) to pianissimo (very soft) and forte (loud) to fortissimo (very loud) with mezzo in between in both cases, but this does not convey the wide, surging dynamics which characterise some pieces, and these early recordings often capture that sense of dynamic drama particularly well (and some composers helped them with descriptive notes on the original scores (Beethoven in particular found the accepted terms insufficient to convey his sense of dynamics and would add his own descriptive instructions accordingly). This is quite surprising, given the changes in technology that we have experienced since then.

If the fine dissection of time and dynamics are important, the other change that one notices is in the attitude of both musicians and conductors. In the early recordings mentioned, the whole orchestra seem to play as one and with an obvious enthusiasm for the music itself. The conductor has, no doubt, played a strong role in this. It is as if the performance is telling a story, and the musicians are eager to tell it correctly and in balance. There is no one musician trying to show off or draw attention from the others. There is a recording of Brahms' Alto Rhapsody, conducted by Bruno Walter, in December 1947. In this recording, which features the wonderful Kathleen Ferrier, everything is in perfect balance and one gets a real sense of drama and romance as you are lead through the story (which was written for the wedding of Robert and Clara Schumann's daughter Julie and is based on one of Goethe's verses). The dynamics are excellent, the choir in balance and, when Ferrier hits those high notes about two thirds of the way in, it really does send shivers down your spine. This is music as it is meant to be. Sensitively played in order to really draw out the story, and *all* music has a story to tell. There have been a handful of recordings of this piece since, but none of them really achieve that special magic that Bruno Walter and Kathleen Ferrier were able to produce. There lays the difference.

Let us consider two currently well thought of musicians on the classical music scene today, a pianist and a violinist. The violinist comes on to the stage dressed in jeans, a T shirt and a mismatched jacket over the top. His hair is drawn back and held with an elastic band, and he is unshaved. Some would hold that this is disrespectful both to the orchestra and to the audience. The author would agree. If the musician concerned pays so little attention to his own appearance, how is he going to do credit to the music he plays? Actually, he is a competent musician in so far as he plays all the right notes in the right order, but there is no feeling in his performance. There is no soul. There is no *humanity*. The pianist is also a competent musician and wants you to note the same. He delights in playing fast runs and gesticulating with both head and arms, exaggerating a feeling for the music which simply is not there. These are 'showcase' musicians who like to show off, but who demonstrate little understanding of the significance of what they are playing. They do not play *with* the orchestra, but within a separate performance of their own. Here, again, is the difference. Really good musicians know how to play together in order to achieve the right sound. It is the same within the orchestra. Some musicians, having secured a safe placing with a major orchestra, simply come in and play what is put on the music stand in front of them. They can sight read and will play all the right notes, but will they *feel* the music? And also with sections of the orchestra, such as cellos or horns. They should sound as one and should be in perfect balance with the overall sound and feel of the piece, helping to *tell the story* that the music is trying to communicate. How often does this really happen with modern orchestras? Occasionally, maybe on a good night. But it used to happen on most of these early recordings. Musicians played differently and were absorbed into the music, and the musical pieces consequently had a different and generally more exciting feel to them.

Perhaps the skills are not being passed on. We do not need 'designer performers' and jobbing musicians. We need people who are really passionate about what they are doing and who are prepared to maintain the traditions of classical music. Similarly with conductors. Those who come on stage wearing some multi-coloured smock are showing similar disrespect for the musicians and audiences. These are the ones who wave their arms about, even when there is no need, and who take every opportunity to grin into the camera, if there is one. It sounds like an unkind comment. It is not meant to be, but surely we need to take a good hard look at how we are playing music and whether we are being true to its intentions. The music should come first. Not the soloist, the conductor or the orchestra. They should all serve the music and the story that lies within it. This is what happens on many of these Polish, Hungarian and Russian State Radio broadcasts. If you were lucky enough to be chosen to be a member of one of these orchestras, you made sure that you played well. And you did not turn up with an elastic band in your hair, unshaven and wearing jeans. In fact, some superb musicians spent some years in these orchestras, and it shows in the overall performance. These might seem like small, unimportant points, but they are not. They are *hugely* important. For they lie at the very psychology of classical music as an art form. The psychology of humanity and civilisation. It is important.

Whether musical instruments themselves have changed is a more challenging question. It is possible that, due to the ageing of both wood and varnishes, instruments have changed slightly in their sound. The construction of strings and bows may have changed slightly in some cases. But these changes would probably be slight and, in any case, overshadowed by the playing technique of individual musicians. In the woodwind section, there may have also been changes in materials and reeds, but whether these made significant changes to the sound of the instruments may be debatable. Clarinetists do tend to favour the wooden instruments over those of composite construction however and, probably, most serious musicians have their own favourite instruments.

There is a movement of course which uses period instruments with gut strings. These are usually replicas of early instruments, built in the same manner. The author has some recordings made this way and the sound is, as expected, quieter and a little smoother. It is hard to say whether early instruments really sounded like this as we do not have access to exactly the same timbers, glues and varnishes. But it is a worthwhile experiment, and of course, there were some different woodwind instruments available in the days of early music. The music still communicates, but with a slightly different voice. And this brings us back to *performance*. Those that played in the 1940s and 1950s seemed to do so with a certain joie de vivre which seems absent today. Orchestras consisting of 49 adequate players are simply not going to sound exciting or to be able to tell the story of music in a unified manner. They need to be *passionate*, well-rehearsed and at one with the conductor. When this happens, you certainly notice it. The author remembers a specific instance of the performance of Carmina Burana at the Albert Hall. It was the last of three consecutive, nightly performances, and everyone was joyous and on top form. The sound was exhilarating and the audience applauded in a manner rarely heard. *This* was how it was supposed to be. We must also look towards education and how music is taught in schools. It should be in such a way as to identify those with a particular aptitude at an early age. Sometimes, one finds the children of musical parents unusually gifted, perhaps because they have been immersed in music right from the start. However, in some countries, the standard of musical education provided in state schools is poor. This is tragic as music is as important as the sciences and, indeed, the two enhance each other. If this is not the case, it means that only those from privileged backgrounds will receive proper musical education. How many must there be who, actually, have a great aptitude and feeling for music, but will never have the opportunity to express it? It follows that the place of proper music in civilsed society needs to be appreciated.

Chapter 21
How to Do Things Properly

It is pertinent perhaps, within this work, to propose some working methodologies which are well proven in the field and which, consequently, can yield good quality, proper stereo recordings. The terms 'proper stereo' and 'true stereo' are used to denote the recording of a three-dimensional sound field with two microphones, as defined in Alan Blumlein's 1931 patent. This area may usefully be divided into three topics: recording techniques, microphone techniques and mastering techniques. If these areas are dealt with properly, then we shall have a good quality recording, whether at a live venue or in the recording studio. If we compromise in any of these areas, then we shall end up with a compromised recording. A classical music performance shall be our example although, actually, the same principles apply to any type of music, be it jazz, folk, traditional or popular.

Much mention has been made within this work of Alan Blumlein's groundbreaking invention of stereophonic recording and playback, way back in 1931. And rightly so. Blumlein was an electronics genius who was many decades ahead of his time and who worked on much else besides, but in creating stereo sound, he left a wonderful gift for music lovers around the world, even if his invention would not be exploited for another 20 years or so.

If we think back to the trajectory of the history of recorded sound, there were many inventions and variations, all aiming to perfect sound transmission and recording, mainly, but not only for audio purposes. Indeed, much of this energy was spent in the field of telecommunications. But, in the field of sound recording, those early pioneers considered themselves lucky if they produced a recorded sound that was even recognisable. By the time we had arrived at shellac discs running at (roughly) 78 rpm and playing on a wind-up gramophone, many felt that the problem had been solved. After all, the gramophone was as ubiquitous as any domestic device could be and the great performers of the day were eager to get their name on a record, which could then be played repeatedly and, conceivably, anywhere in the world, thus providing them with a tremendous and previously unheard of exposure. Discs

© The Author(s), under exclusive license to Springer Nature Switzerland AG 2021
J. Ashbourn, *Audio Technology, Music, and Media*,
https://doi.org/10.1007/978-3-030-62429-3_21

were relatively inexpensive, and the gramophone became the centrepiece in house-holds everywhere. Everyone was happy.

Had it not been for Blumlein, tape recorders would have come along and the quality of recorded sound would have improved, but it would have no doubt remained in monophonic format, even though, in the film world, people had started to look at the possibility of multi-channel audio, but this is a different thing entirely. The wonderful thing about true stereo sound is that, when undertaken correctly, a tangible, three-dimensional sound field is created, within which, performers are placed accurately in space. Furthermore, this space contains width, depth and height information. Consequently, if performers move across a stage, as would happen within an opera for example, their movement is accurately tracked. If they move to the back of the stage, this will also be accurately depicted within the stereophonic field. Blumlein himself made a short film showing movement across a stage and back again, all reproduced beautifully in stereo. It is important to understand the distinction between two-channel audio and stereo. Two-channel audio is exactly what the name implies, audio recorded onto two channels of recordable media. It might as well be three channels, or six channels, or 24 channels. The sound remains as monophonic signals recorded to multiple channels. Stereophonic sound is com-pletely different. It is a methodology, using an understanding of physics, with which to create a fully dimensional sound field and to be able to record that sound field in such a manner that it could be reproduced after the event. The fact that Blumlein managed to reduce this down to just two channels is astonishing. Bear in mind also that there were no stereo playback devices. So Blumlein also invented the stereo-phonic record, using both height modulation and side to side modulation in order to capture the two channels. He even designed cutting lathes with which to manufac-ture stereo discs.

The interesting thing is, that if you undertake a true stereo recording, as defined by Blumlein in 1931, it works. You get a wonderful, three-dimensional field with everything placed correctly within this field. In addition, stereo recording accurately captures the acoustic in which the recording takes place, as it is also capturing the natural reverberation taking place, between instruments and between instruments and the space in which they are being played. Consequently, it is ideal for recording orchestral performance, operas, chamber works, choral works, traditional music and just about anything. And, to accomplish all of this, all you need is two microphones and something capable of recording their output.

The reason this was such a revolutionary idea is that Blumlein, being a good engineer, went right back to first principles in order to understand how sound works. How sound waves behave in enclosed spaces and, especially, how we hear sound with two ears on either side of a head. This last point is crucial, for Blumlein realised that it is time and phase differences, as received at our ears, which enables us to determine direction and relative loudness of the variable sounds that we hear at any one time, with our two ears. He was thus determined to work with just two micro-phones and two channels. Bear in mind also that the available microphones at the time were rather crude devices (and so Blumlein also designed the moving coil microphone) which made experimentation difficult. Nevertheless, the invention of stereophonic sound was perfected. It worked then, and it works now. The problem is that relatively few audio engineers record in true stereo. Perhaps they are afraid to reduce everything down to a single pair of microphones. Perhaps they do not know what they would do with all those additional tracks. Perhaps they simply do not understand how stereophonic sound recording actually works.

The first step then, along the road towards doing things properly, is to record in true stereo using either a coincident or near coincident pair of identical microphones. A favoured position in which to place them is above the conductor's head and angled so as to point towards the middle of the orchestra. This is certainly a very good starting point, although final, precise positioning should be made by ear. The reason that this represents a good starting point is that the conductor has expended a great deal of effort, through rehearsals, in order to obtain the sound that he is trying to get at the podium position. If everything sounds just right to him on the podium, then that is surely a good place from which to record. A couple of feet above his head and angled towards the middle of the orchestra should obtain a fairly balanced sound, *if* the orchestra is laid out in a conventional fashion with plenty of space around them. However, sometimes this is not the case, due to the architecture of the venue, and the orchestra might have been deployed in a non-typical manner. This is where an experienced audio engineer will think on his feet and come up with an innovative solution. Such a solution might involve the deployment of a second pair of microphones in order to ensure that everything is covered but, in any event, the main sound will be that picked up by the primary stereo pair.

An interesting way of checking coverage within a non-standard orchestral setting is for the engineer to do so with their eyes. Knowing the polar pattern of the microphones being used, it is not difficult to look along their length and visualise the field of acceptance that is likely to be captured. If there is a section, such as a choir, which is clearly outside of this field of capture, then another pair of microphones might usefully be deployed and a little of their output mixed in with that of the main pair in post-production. In any event, we start out with a true stereo recording. It is a good start.

Now the question is which microphones to use, and in which configuration? The 'coincident pair' configuration, also sometimes called a 'Blumlein pair' depending on which type of microphone is used, is a good starting point. The coincident pair technique uses two cardioid 'pencil' microphones, arranged at 90° to each other and with their capsules in the same vertical plane, one above the other. This technique is easy to set up and has several advantages. Firstly, a true stereophonic field is created, with excellent spatial information and a good relationship between stereo width, depth and height. The resulting sound is detailed and coherent with a strong central image. Furthermore, it has excellent compatibility with monophonic sound, allowing a broadcast radio station to play the recording over a monophonic, or stereophonic channel, with both sounding excellent. If, instead of cardioid microphones, those with a figure of eight polar pattern are used, then the sound field will be almost 360° and a wonderful sense of the acoustic will be provided (together with any audience noise which may be seen either as an advantage or disadvantage). This is what is known as a 'Blumlein pair'. Figure of eight microphone polar patterns often have a dominant lobe and a smaller lobe. Naturally, the dominant lobes should be the forward facing pair. In any event, it is remarkable to be able to capture so much information from just two microphones. Undertaken properly, this technique works and provides excellent results (Fig. 21.1).

As always, there are variations on a theme and audio engineers started to experiment with near coincident pairs of cardioid microphones. This configuration is similar to a true coincident pair, but the capsules are not aligned in the vertical plane. This means that there is not such a strong, phase coherent central image, but there is a wider stereo image. If the orchestra has plenty of space and has been laid out in the conventional manner, and the audio engineer is placing his microphones in close proximity, for example, above the conductor's head, then this wider stereo image at the expense of a slightly less strong central image may represent a good compromise and provide a wide, spacious sound which remains highly detailed and, if the audio engineer has set his levels properly, will capture a full dynamic range. The popular near coincident pair techniques include the German DIN specification with the capsules spaced 20 cm apart and the microphones angled at 70°, the Dutch NOS specification with the microphones 30 cm apart and angled at 90° and the French ORTF specification with the capsules 17 cm apart and the microphones angled at 110°. For the ORTF specification, it is possible to buy stereo microphones (in one body) with the capsules already arranged properly. The use of these microphones makes for a very fast and tidy set up, while almost guaranteeing a good quality recording, assuming that the engineer knows what he is doing (Fig. 21.2).

There are other possibilities. Some audio engineers favour a simple array of two or, sometimes, three microphones, placed in a line in front of the orchestra. There are also various suggested configurations for a pair of microphones, quite widely

Fig. 21.1 The Blumlein coincident pair pickup pattern

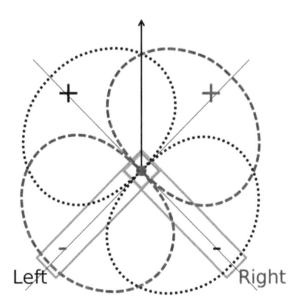

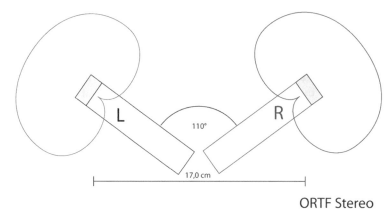

ORTF Stereo

Fig. 21.2 The ORTF near coincident configuration

spaced, in front of the orchestra although, such arrays can produce phase anomalies if not undertaken very carefully.

In general, it is recommended that audio engineers experiment with the near coincident pair specifications, using good quality cardioid microphones. Align them visually to get the right angle of slope above, and maybe slightly behind, the head of the conductor, pointing them towards the centre of the deployed orchestra. There may be exceptional circumstances, such as when a choir is deployed towards the rear of the orchestra, when a second stereo pair of microphones might usefully be deployed, or even two spot microphones. This is a technique that the audio engineer may hold in reserve in order to cover unexpected or otherwise exceptional situations.

The benefit of using either a coincident or near coincident pair of microphones is that the resulting sound will be true stereo, with a tangible, three-dimensional sound field established, with the performers in their correct places within the spatial domain. Furthermore, with excellent width, depth and height information also being captured, the recording will have a strong sense of 'being there' for the listener. If figure of eight microphones are used, that sense will be heightened, with the listener feeling that he or she is actually sitting in the audience. But, most of all, the orchestra will sound like an *orchestra* and not like 47 different instruments mixed together. This is very important for seasoned music lovers and musicologists who wish to study the performance and sound of a given orchestra playing a particular piece. And so, the way to go when recording a classical orchestra, a choir, a chamber group or a small group playing or singing traditional music from a given region is most definitely to record them in true stereophonic sound. Adopting this approach also allows the audio engineer to remain as unobtrusive as possible and to keep out of the way of the musicians. This is desirable, both at rehearsals and the live event, and also holds true within the recording studio.

The last step in the process of making a good recording is, what is often referred to as 'mastering' or sometimes 'post production'. What this really means is that the audio engineer has now returned home with his recordings and is ready to make the final, two track, stereo recording. This may then be used to manufacture CDs or perhaps provided online as a high-resolution file (if it has been recorded in high resolution). There are all manner of special effects and techniques that may be deployed at this stage. More than you can shake a stick at. The author's view? Do as little to the original file as possible.

There seems to be a misconception that editing in the digital domain is entirely transparent and cannot possibly cause distortion, or soften the signal due to multiple passes and adjustments to the original file. Well, if you think about it, this cannot be true. If you do anything at all to an original file, such as normalise it, add equalisation or compression, etc., then you are altering the numbers which represent each sample. This is a form of distortion as you are moving away from high fidelity to the original. And so, bit stream 'A' for a short burst of music becomes bit stream 'A+M' where 'M' is the mangled element of the sound. If you inflict 'M' multiple times upon a section of music, the numbers representing the samples within this section will now all be different. The sound will bear some resemblance to the original, naturally, but something will be lost. In addition, the inherent phase and spatial information will be affected. Usually, this is most apparent upon transients and fine detail. A good axiom for new audio engineers might be 'MAD'. That is, mangling adds distortion.

However, the files might need some work. For example, the engineer will wish to ensure that there is a nice clean 'start' and 'end' to each movement. Consequently, he will be careful to remove any noise from these parts of the signal, creating silence, but not with an abrupt cut-off point. The low noise floor should blend seam-

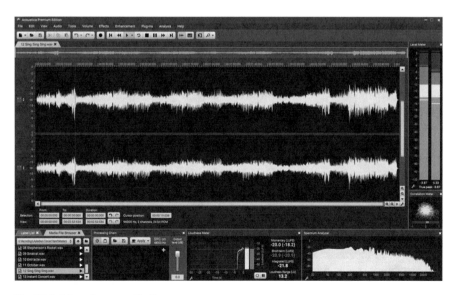

Fig. 21.3 Mastering an audio file

lessly with the start of the signal. Similarly with the end. The engineer might also remove any dc offset and, if down-sampling for CD, may choose to add a little dither. Other than this, if he or she obtained a good quality recording to start with, there should be little or nothing else to do. An interesting test that one can make of this reality is to go ahead and modify your file with equalisation (tonal correction), compression, add a limiter and some artificial reverberation from an expensive programme. When complete, compare it to your original backup file and see which one sounds more lifelike, more immediate, more cohesive—in short, better. Ah! Now you can hear how desirable it is not to become a MAD engineer but to remain a disciple of the truth, in audio terms. Taking this last route is very satisfying and, after a while, one develops an intuitive feel for microphone placement, ensuring that the original file will always sound clear, with good dynamics and fine detail.

Another advantage of developing a good microphone and recording technique is that there will be very little to do at the mastering stage. In Fig. 21.3, one can see visually that a nice clean stereo file has been captured, with good dynamic range, and from the spectrum analysis at the bottom right that a full frequency sound has been captured and from the correlation display above that it is phase coherent. Note, that there is energy here, right up to and beyond 20 kHz, but it is starting to fall away as it approaches this point. This is a recording of a medium-sized concert band, made with a near coincident pair of cardioid microphones.

In short, record always in true stereo, develop a good microphone technique and do as little to the original file as possible. Then you shall be making good recordings which sound immediate, clear and detailed. With practice, this becomes a natural process and is easily deployed in most situations.

Chapter 22
The Use of Digital Audio Workstations and the Impact on Music

Once it was realised that you could store digital audio on a computer, why not use the computer to record it in the first place? This was the time of the word 'workstation' when applied to any computer undertaking a specialist task, and so the Digital Audio Workstation was born. Manufacturers had been tinkering with this idea since the late 1970s (Soundstream), although computers were not really up to the task at that time. In 1983, the MIDI (Musical Instrument Digital Interface) standard was introduced, which allowed devices to be connected together and played in synchronisation. Ideas were bouncing around, and it was a fertile time for suggestions, but it was not until 1989 that we saw the introduction of sound tools from Digidesign, and this was really the first practical product, based on the Mac computer. In 1991, they renamed it as Pro Tools, and it had non-destructive editing which was seen as a huge benefit. Producers and audio engineers could try all sorts of ideas without destroying the original sound file. In 1992, Steinberg responded with Cubase Audio Mac, which incorporated MIDI and also had a music notation capability.

One could go on and on, listing a dozen or more products which all followed on in a similar fashion. The seed had been sown and everyone, including established studios with their big tape machines and mixers, wanted computer functionality. Hobbyists also quickly realised that they could, for the first time, use the same technology as mainstream studios. Anyone could buy a copy of Pro Tools or Cubase or any of the other systems that were cropping up. All you needed was an interface between the computer and your microphones or electronic instrument, and they too were starting to appear at reasonable prices.

At first, the professional world saw computer audio as an interesting toy. However, slowly but inexorably, the toy took over and everyone wanted to use it in order to produce their albums—at least in the popular music field. In the classical music field, things were rather different. Here, specialist digital recorders were designed for live recording, with special attention paid to sound quality. While the DAW was designed to build up a track, layer by layer, sound by sound, the classical world just wanted to record in digital. However, the capability to edit within the

J. Ashbourn, *Audio Technology, Music, and Media*,
https://doi.org/10.1007/978-3-030-62429-3_22

digital domain, and to a fine degree of granularity, was welcomed. However, in this chapter it is the DAW which shall be considered, together with its impact on the popular music scene.

Today, anyone with a computer can make music. Just acquire a DAW (some of them are free) and some sampled instruments or sounds and off you go. If you have a MIDI keyboard that will connect via a USB socket on your computer, and with that, you may play any of the sampled instruments and make professional sounding tracks (assuming you can play to at least a basic standard). Steinberg had previously introduced the Virtual Studio Technology (VST) standard, which enabled sampled instruments or special effects to easily be loaded into the DAW of your choice.

And then, they came, one after another. Everyone who could write software seemed to come up with a DAW. Some of them were buggy and atrocious, but those that were not too bad evolved and became more reliable and, of course, with more features. Nowadays we have around a dozen or so leading examples, all of which, quite honestly, are very similar. The author has experimented with most of them and found that they all do more or less the same thing, more or less the same way. There are two of them however which stand out for doing things slightly differently.

Tracktion Audio implemented their first versions with a single-screen interface, wherein everything could be seen on one screen, including the mixer settings. They remain faithful to this layout (although they now include a separate mixer screen) and one can imagine that users, having worked this way, would also remain faithful to the product. It incorporates many other good ideas that will appeal to those who like working with audio. Another unique feature of Tracktion Audio is that, when a new version of their software is announced, they make the outgoing version freely available. This is a good marketing idea which has, no doubt, brought many young customers their way (Fig. 22.1).

The other DAW which takes a slightly different approach is the Harrison Mixbus. The Harrison company became very well known for their physical audio mixers which have been used extensively in film and TV studios. They had a particular sound which these producers liked. When they decided to market a DAW, they took the unusual approach of emulating their own mixer, right down to getting that particular Harrison sound that everyone liked. Indeed, this DAW really does sound different and also has a layout which will be familiar to studio engineers using these large mixers. Consequently, they will feel immediately at home with the operational logic of the Harrison Mixbus product. It is a good marketing approach (Fig. 22.2).

The operational methodology of most of the DAWs is very similar. One records individual tracks in a track view, which shows the recorded tracks one on top of the other, and then, when ready, one can switch to a mixer view which allows for a more intuitive way of mixing the tracks down to a two track master. Many of them allow for portions within a track to be arranged in a stack and played back in that order. Consequently, one could take a DAW on stage and play an entire work without ever touching a real instrument. Similarly, disc jockeys could use a laptop computer with a DAW and a collection of sounds and tracks which they could arrange on the fly and play back in real time. And so, the DAW itself has evolved to do much more

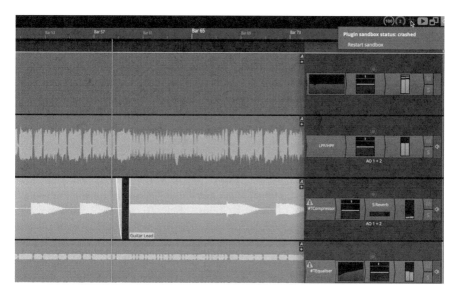

Fig. 22.1 Traction Audio's waveform DAW

Fig. 22.2 The Harrison Mixbus DAW

than just record tracks. However, it is to record tracks which is their main purpose and some make this much easier for the user than others.

It is fair to say that all DAWs have a steep learning curve if you are new to them. Even if you have worked previously with tape and know your way around a recording studio, you will find some things strange at first. In addition, they are all idiosyncratic, and if you become really familiar with one of them, you will still struggle to pick up the working methodology of another. This is how things are.

The question is, what do they sound like? It depends on how they are being used. If they are being used like a multi-track tape recorder to record live sound, then, if you record in high resolution and with a good quality digital interface or mixer, then they can sound very good indeed. The main component of interest in this case will be the analogue to digital convertor and where it resides. If the ADC is hardware based and is within a high-quality interface or mixer, then the sound that it delivers to the DAW will, no doubt, be of good quality. If it is software based and is in your computer, it may be variable. In any event, it will not sound bad.

Of course, you may decide not to record anything with a microphone but to simply used sampled instruments, played from a MIDI keyboard. In such a case, much will depend upon the quality of the sampled instruments, some of which are very good indeed, provided they are played within their natural range. But, of course, such recordings will still sound artificial as there is no natural acoustic being captured. Nevertheless, this is how a good deal of music, especially music for film, is being made these days. Even established stars within the popular music field are becoming used to coming into the studio and simply singing over backing tracks which have been made with a DAW. Some DAW manufacturers even have 'accredited engineers' who are experts in using their particular DAW.

But using these software programmes is not necessarily easy, and many will struggle to find their way around them. From Fig. 22.3, one can see that, even on a fairly large computer screen, the controls themselves can be quite small. Here we can see that some tracks are currently active as indicated by the channel strip 'meter', and on the extreme right of the picture, we can see the master two-track

Fig. 22.3 A typical DAW in mixer view

mix. Next to this indicator is a 'return' track, in this instance from a reverberation software device. One might have several return tracks in some instances. The 'send' controls for the return track may be seen above the level indicators. And so, these software DAWs, for the large part, emulate hardware mixers. Except that they are not tactile. Everything has to be adjusted with the computer mouse, or shortcut key commands that no one can remember. If one had never used a physical mixer or multi-track tape recorder, then, of course, it would all be new anyway, and so it might be easier to absorb the working practice of these devices. A simple factor such as tempo change, for example, can become quite complicated with these devices. There is always automation, of one sort or another, but this is often undertaken in a cumbersome manner and can lead to unexpected results. The DAWs currently on the market are all quite powerful and usually offer many tracks and facilities. However if you really expect to use all of these facilities, then you will be in for a shock.

First of all, you may have 256 or more tracks available to you, but can you really imagine what 256 tracks are like, to administer via a computer screen? Figure 22.4 shows a DAW in track view where just ten tracks are being used and, already, the screen is full from top to bottom. So if you were alternating between edits on track 3 and track 29, you will be scrolling up and down like mad on your computer screen. Similarly within the time domain. Figure 22.4 shows just a few minutes of sound. Supposing you are working on a piece that is 20 min long? You will need to keep shuffling back and forth on the screen, or zoom out to such a degree that you cannot easily identify where you are.

Obviously, if you are going to work properly with a DAW, then a very large, high-resolution computer screen is a prerequisite. There are two other requirements.

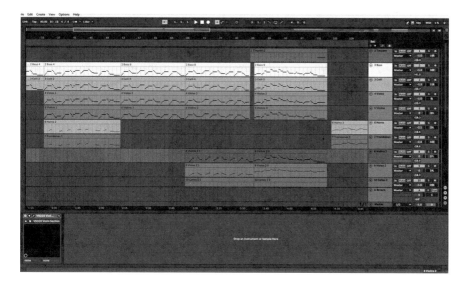

Fig. 22.4   A DAW in track view

You shall need a *very* powerful computer processor and a great deal of memory. Every time you play a note, you are asking the computer processor to play a sample from a programme that is on your hard disc. It has to know exactly where this programme (the sampled instrument) is, the note you have requested, and how you want it played. But there may be another 39 notes, all being played at that moment in time, and the processor needs to undertake all of them. It must also manage any effects such as reverberation, compression, delay and so on, all in real time. On top of all that, it must be running the DAW software itself and it has to run the operating system. As you might imagine, all of this requires processing power, and the more you have running, the more power you require. If you have a little laptop that you bought in a sale 10 years ago and think you are going to emulate a 256-track recording studio on it via a DAW (which is bloated anyway), then you shall be in for a surprise. If you really want to run a DAW properly, then it will be better to custom build your computer. Choose a *fast* current model motherboard with good inputs and outputs, and invest in the latest and fastest processor that you can afford. Also, buy a large, slow-running cooling fan with a large heat sink to place on the processor. Think of how much memory you can afford, and then double that amount. Sixteen gigabytes might be reasonable, but more will be better. Choose the *fastest* hard disc that you can possibly find, with the greatest capacity. If you choose to use solid state storage, you will need a lot of it and this will become a major expense. A hybrid drive with enough solid state capacity to run the operating system from memory might be a good choice. And then place all of this in a rugged chassis that can be placed under your desk and largely forgotten, except when you are undertaking regular backups.

With a computer such as described, you will probably be able to run as many tracks as you actually need (which might be no more than 16) and run them well. The problem of latency often occurs when you are adding tracks, especially when using an external computer interface. This is simply the slight time delay between the tracks playing back and your new input, by the time it has made its way through all the circuitry. This is another reason that you need that fast, expensive processor. Of course, if you are mainly recording yourself playing instruments, you could forget DAWs altogether and buy one of the many 'portastudios' on the market. These are of very good quality and retain the tactile feel of mixers. They also seem to offer astonishing value for money, presumably a result of surface mount technology upon a single board. The top models are very flexible and can perform most functions, including on board effects such as reverberation. In addition, there are conventional mixers, with all the usual facilities expected on such a device, but with the added capability of recording the mix on an integral SD memory card. A great idea for musicians and small studios.

Whether we like it or not, the digital audio workstation has had an enormous impact upon recorded music. Within the popular music field, it has enabled two minor revolutions. Firstly, within the conventional studio, it is changing the way they work. Nobody uses tape anymore, everything is accomplished with computers. This allows for easier editing and more easily allows for special effects to be integrated with the music. Secondly, the arrival of the DAW has meant that the teenager with aspirations to be a pop star can have the same equipment in his or her bedroom as the major studios have. They will certainly be able to produce quite a good quality of sound, regardless of the quality of the music. They will then be able to compress this down to a smaller file size and place it on the Internet for others to access, probably all in the same day. This reality has brought us into the age of instant music. There is much more of it than we can possibly consume and, sadly for those growing up now, much of it is of quite poor quality. That is not just a personal view. It is a matter of repeatable scientific assessment. A piece of music which has no structure, no melody and no harmonies within it, cannot really be considered as music in the way that civilisation understands it. Not all modern music is like this, of course not. There are, buried among the majority, some quite nice songs, the problem is finding them.

In the classical world, things are a little different. Similar DAW software may be used but often with specialist, hardware based controllers. If, as an audio engineer, you intend to record in true stereo, with maybe the exceptional second pair of microphones, then you do not need dozens of inputs. Six or eight inputs will probably be enough. Consequently, there are several approaches one could take to recording a live concert. Indeed, one approach is not to use software at all, but simply use one of the good quality six- or eight-track recorders which are currently available. These have the advantage of being extremely portable and offer high-resolution recording. There is much discussion about the quality of their microphone preamplifiers and analogue to digital convertors, but the better models are fine in this respect. Coupled with good quality microphones, these little devices are capable of making acceptable quality recordings.

The next step would be to use a good quality digital interface coupled to a laptop computer. This interface may have all the microphone inputs you need and, other than level control directly on this interface, you might be happy to do everything else within the DAW software on the laptop. After all, any editing will typically be undertaken after the event in post-production. This approach is similar to using a portable tape recorder.

There are also custom made, and sometimes quite expensive, control surfaces, containing the necessary microphone inputs and convertors. These will typically provide level sliders and other controls and will be much easier and more intuitive to work with in the live environment. They will typically include automation and use a laptop to run their own software and offer a display of the waveforms being recorded. This is a good compromise for live work as a meaningful hardware interface is provided in a compact package that can simply work with almost any (fast) laptop. The whole thing remains transportable for the audio engineer and, yet, can provide excellent quality.

Other variations include splitting out the hard drives into special enclosures which allow for mirroring and redundancy. This could be important when recording an event which will not be repeated again. Of course, in such a situation, the audio engineer will also use redundant power supplies and other fail safe measures. But the software will still resemble the DAW.

Suffice to say that the DAW has had a tremendous impact upon the recording industry and has virtually banished tape, although some studios have been wise enough to hold on to their multi-track tape recorders as some clients love them. The ease of non-destructive editing within software has become accepted and few audio engineers learning their craft today will go anywhere near tape (which is a shame).

There exists another option, as briefly mentioned above, for live recording where absolute quality is not required. It was obvious that, sooner or later, someone would think of taking a transportable mixer and building in the functions of the portable digital recorder. The result is a nice, conventional mixer, with typically up to 16 or 24 channels, complete with phantom power on the microphone inputs, and a simple SD card digital recorder, usually with up to 24-track recording, depending on the resolution chosen. This approach is perfect for touring bands who wish to control their sound and record their performances. The visual interface of the DAW is absent, but who needs it if you have a proper mixer? A slightly up-market version of one of these devices, recording on two cards (mirroring), with super low noise microphone inputs and good quality analogue to digital conversion might present a good option for recording classical concerts. Furthermore, you would not really need 24 tracks, so smaller, more portable devices could be made.

However, it looks like the DAW is here to stay, and it is clear that one problem is already being faced by the manufacturers, and that is how to cram in even more features for the next version of their software. Already, most of these software packages are so bloated that they will take over the computer they are loaded onto. Most of this bloat comes from quite useless 'sounds' that they, the software supplier, think you should be using. This explains why so many popular music offerings sound so similar. Even if you deleted all of these sounds, you are left with a thousand and one mindless features that you will probably never use, or even know about. And yet, they still miss some of the obvious requirements. It would be nice to see a truly lightweight DAW which is focused on usability, enabling the musician to produce individual music which, preferably, may be replicated in the live environment.

DAWs have enabled multiple tracks of music to be constructed, layer by layer, to whatever lengths the operator wishes. In most cases, the limitation is computer power and memory. Several software packages offer an unlimited number of tracks, which is quite unnecessary. Some offer up to 256 tracks, which would be unusable on most computers. Hardware multi-track recorders using disc arrays typically offer up to 64 tracks, but few would exploit even this capability. The other things that DAWs do is eat up time. Gone are the days when a popular artist would go into the studio and record a song in one take. Now they will require weeks. But classical concerts are usually recorded in one take (sometimes the audio engineer will also record the dress rehearsal) and, even classical recordings in the studio, if arranged properly, may be undertaken quite quickly.

There is a lovely EMI box set of their recordings of the late pianist Solomon, undertaken mostly in the 1950s. It is instructive to read the accompanying notes. Ah! Now we can see the impact that computers and DAWs have *really* had in the recording world. And not just for recording, but for performers as well. How many pianists today would be happy to record Beethoven's fifth piano concerto in one take? And get it right? Or his violin concerto? Tape allowed them to splice different takes together, within reason, but computers allow them to focus in on individual notes, to alter the tempo at will, and generally play about with it to their heart's content. This capability is moving away from what music is really about. We are bleaching the spontaneity from it and lowering the overall musical capability. This is not a good thing.

Chapter 23
Why Recordings Sound Worse Now Than They Did in the 1950s and 1960s

Many would raise their eyebrows in alarm at such a suggestion but, in many cases, it is absolutely true. One only has to listen to some of the classical recordings that were made by companies such as Decca, EMI and DG from this period in order to recognise this reality. Bearing in mind that a 'record' is supposed to be just that a record of an event in time. An audio record being a record of the sound of a specific event in time. If you made a written record of an event in your diary, you would record the event accurately, with as much detail as possible, in order that you may recall it accurately at a later date. And so it should be with an audio record. And, in many cases, this is exactly what audio records were. Audio engineers were sent out to a specific venue to record an event taking place, and the producers wanted the result to sound as though the listener were present at the time. Similarly with broadcast concerts and, for a while, similarly with studio recordings.

The primary reason that these early recordings sounded so good was because they were recorded in true stereophonic sound, just as Blumlein had defined it. In addition, they had been recorded with simple but good quality equipment. Imagine the signal from a pair of high-quality microphones feeding straight into the preamplifiers of a good, two-track tape recorder running at a minimum of 15 inches per second. The sound will be phase coherent and very clear with excellent transient response. This is why these old tapes when cleaned up now to remove tape hiss (which was very low anyway) sound so good. Actually, even monophonic recordings from around this time sound good, and with some, depending on where they placed the microphone, there is quite good depth information, clearly separating the orchestra, chorus and soloists in space. But it is the proper, true stereophonic recordings which really sound good and give a sense of being there, sitting right in front of the orchestra.

These true stereo recordings were made mostly in England, followed quickly by Europe and then Russia and the eastern European communist countries. In America, things were a little different and the definition of stereo seemed to get confused with two channels. Very often, recordings were made with an array of two, three or

J. Ashbourn, *Audio Technology, Music, and Media*,
https://doi.org/10.1007/978-3-030-62429-3_23

sometimes four microphones, simply placed in front of the orchestra. This did provide an exciting sound, with a degree of correct instrument placement, but it was not
stereo and, consequently, did not exhibit the wonderful, three-dimensional field that
stereophonic recording provides. Some American audio engineers went from this
approach to multi-track tape recording when that was introduced, never really
understanding the benefits of true stereo.

Back in Europe, radio broadcasts were starting to be recorded for archive purposes, and how lucky we are that this was the case. If you had an orchestra in the
radio studio, or you had broadcast engineers out at the live event, having got everything set up and ready to go on air, it was a relatively simple thing to take a feed out
from the broadcast desk, straight into a high quality tape recorder (or maybe two if
the programme was longer than a reel of tape allowed). The signal route was very
straightforward and the signal received at the tape head was clear, phase coherent
and with nothing blocking the transients. This is why these recordings sound so
good today.

As multi-track tape recorders came along, some audio engineers still recorded in
stereo but appreciated the facility, should it be required, to add another microphone
or two to specific areas, such as choirs set back in space, in order to have the facility
to reinforce that particular sound, should it be necessary to do so. Other engineers,
unfortunately, started to put out as many microphones as there were tape tracks,
with microphones in front of every section and, in some cases, almost every instrument. They would then record the signal from these microphones at an optimal level
and, in post-production, mix all of these signals together into what *they* thought the
orchestra should sound like. The immediate result of this approach is that orchestras
no longer sounded like orchestras, but like an assemblage of sounds with a completely artificial balance. The mixers they were now using with these multi-track
tape recorders had a channel 'strip' for each track on the tape. On this strip would
be a master gain control at the top, a selection of equalisation or 'tone' controls, one
or more 'sends' that could send the signal to an external processor, a level faded
with which the engineer could adjust the level and a 'pan' control with which the
engineer could place the signal anywhere between the two tracks. The engineer
could then decide exactly where the instrument appeared, how loud it would be in
relation to the other instruments, he could adjust the tonal qualities or timbre of the
instrument and also add effects to the sound. And he could do this for every channel
or track from the tape recorder. Unsurprisingly, the result was a mess. Yes, of course,
Mozart still sounded like Mozart, but not how you would have heard him at the time
and not how you would hear him now, at a live concert.

The multi-track approach had its followers, partly because it was more forgiving
of slight mistakes. Partly because, with concertos especially, a solo instrument could
be lifted right up in the mix, enabling every note to be clearly heard. And partly
because it made so many adjustments to the sound possible after the event. Nobody
had to think too much about what they were doing during the recording. Just put
microphones up everywhere, set the levels so as to get a good signal and then sit
back and relax.

Fortunately there were, especially in Europe, a handful of good audio engineers who realised that this was not the way to make good recordings. They stuck to their stereo guns and the skills that they had learned the hard way about microphone placement. But they are the minority. Most of the classical recordings from most of the big record companies that you buy today are recorded with the multi-microphone multi-track approach. What is wrong with that? One might say. After all, you do get a nice, crisp, squeaky clean sound to play on your hi-fi system or in your car. Indeed, but it does not sound like an *orchestra* playing an orchestral work. It sounds like an assemblage of instruments thrown together, brought to the back or front as the piece progresses and panned arbitrarily from left to right. This is *not* high fidelity. All the complex phase and spatial cues have been completely lost and the sense of a natural acoustic completely shattered. Even worse, very often an artificial reverberation has been added, and this cannot possibly work as the instrumentalists were never in that artificial reverberant field that is added on top of the whole mess.

If you stand next to the podium while an orchestra is playing one of the wonderful works that have been left to us, the sound is quite thrilling in its dynamic and its overall spatial domain, partly a signature of the concert hall. Furthermore, instruments play at the level they are supposed to, as indicated by the composer. If a flute solo sounds quieter than the full orchestra, then that it is how it is supposed to be. Furthermore the width, depth and height of the sound stage in front of you is all clearly delineated with every sound from every instrument, exactly where it should be. The overall effect is quite wonderful. Stereophonic recording preserves all of this, exactly as it should be. Multi-track recording does not. This is precisely why many modern recordings may sound bright and clean, but they are not proper records of the event. Simply an artificial facsimile. A collection of sounds mixed together and presented to you as the audio engineer and producer wish to (but not how Beethoven would have wished to).

Those who doubt this should look for some early stereo recordings and listen to the difference. Very often, you will find yourself absorbed by the music itself, which is how it should be. And, if you are listening critically to different recordings of the same piece, you can only really do this with true stereophonic recordings. This is why we should promote stereo at every opportunity and make sure that we do not lose the skill of proper stereophonic microphone placement and recording.

In the field of popular music, there were some quite good quality monophonic recordings made. This was probably because the more popular stars were signed to the big labels who had proper recording studios and experienced producers. However, as soon as multi-track came along, this all changed quite quickly. There were very few popular music records made in stereophonic sound. That is a shame, as one could have made some interesting recordings, especially with traditional and folk music using acoustic instruments. Many music lovers today have never heard true stereo and, similarly, have never heard an audio signal that has been recorded 'raw', that is, without added effects. That is a tragedy as far as the arts in general are concerned. We need to take a step backwards and re-think how we record a musical event. The primary objective should be accuracy. We can easily do this. We have some good tools at our disposal, but audio engineers need to be trained more in the appreciation of music and less in the appreciation of software. Manufacturers have a part to play here as well.

Chapter 24
Music and Civilisation: Why It Is Important

At some time in human evolution, man discovered that he could make sounds purely for pleasure. He no doubt experimented with the sounds he could make with his own voice, as well as the sounds he could make with various natural objects. Finding a broken bamboo shoot or something similar, it would have been instinctive to blow into it and see what happened. At some stage, no doubt, blowing into things and banging other things together made an accompaniment to the sound of the human voice or perhaps multiple voices. Children would, no doubt, have happily joined in on these occasions of making sounds purely for pleasure. And there would have been the musical sounds of the forest around them as birds and animals communicated with each other. It was a wonderful world of sound.

We have little archaeological evidence from the Mesolithic period, but in the Neolithic period, there is plenty of evidence. The ancient Egyptians enjoyed singing and dancing and had a wide variety of instruments. Flutes and other woodwind instruments, percussion instruments, and both fretted and un-fretted stringed instruments, including harps, were in evidence, and it is obvious that celebration with music was an important part of ancient Egyptian life. Unfortunately, we do not know what their music sounded like, but we do have poetry which often formed the song lyrics. In the middle kingdom especially, their music, like everything else, would have flourished under the pharaoh Senruset (actually, there were three Senrusets, alternating with another family, to steer things along). The wonderful city of Abdu (which is buried under the present day Abydos) would, no doubt, have had areas set aside for the performance of music. They clearly understood the importance of music to civilisation. It not only reflected the wonderful times in which they lived, but it gave them enjoyment on several levels, from joyful song and dance to more sophisticated music which would have been intellectually stimulating.

Jumping forwards to Tudor times, music at court and at celebrations was very important. Henry was himself an accomplished musician and a great patron of the arts (in fact, of education in general). He not only wrote music himself, but encouraged others to do so, and we have plenty of music from this period. The lute was a

J. Ashbourn, *Audio Technology, Music, and Media*,
https://doi.org/10.1007/978-3-030-62429-3_24

popular instrument, with several variations as to the number of strings and the shape and size of the sound box. John Dowland was a prolific writer of music for the lute, much of which has survived in notation form, from which we can appreciate that it was very sophisticated music. Choral music was equally important in Tudor times, with hundreds of pieces written for both the church and special occasions. Composers William Byrd and Thomas Tallis managed to get a monopoly on music publishing, and both flourished under the Tudors. It is interesting that William Byrd was a Catholic throughout his life, and yet, somehow, this was accepted by the Protestant Tudors, perhaps because they knew good music when they heard it. Byrd was a master of polyphony, and his masses in particular were intricate works of art which reflected their own times. Music was important to the Tudors. Elizabeth was also an accomplished musician and continued in her father's footsteps as a patron of the arts, both musical and visual. The sophistication of English music at this time reflected the scholarly nature of the court and, indeed, important houses and families throughout the land.

Moving to the wider European period, Telemann, Purcell, Bach Handel, Haydn, Mozart and, especially, Beethoven moved music forwards in leaps and bounds as life within the European courts became increasingly sophisticated. Great art was also created in this period, and architecture adopted a grand and wonderful style. But, in music, it was Beethoven who really reached out and broke boundaries in every style of music for which he wrote, whether it be piano concertos, symphonies, masses, string quartets or for some of the more unusual combinations for which he wrote such wonderful music. Beethoven knew and admired the work of Byron, Goethe, Schiller and other poets and philosophers. He loved to read, and he loved the countryside and nature. His work evolved into something much more than just court entertainment, and many hold that Beethoven represents the peak of civilisation. His later string quartets took music in a different, more cerebral direction, and his masses communicate much more than the expected. Both the Mass in C and the wonderful Missa Solemnis are a revelation when compared to anything that had been written up to that time. The Missa Solemnis is full of pure Beethoven, even though it uses a popular framework. Equally, the fifth Piano Concerto reflects almost every human emotion and more. Yes, these works surely represent the peak. Interestingly, Beethoven was convinced of the healing powers of music and his famed improvisations often reflected this, especially if played for someone he was fond of. His music also reflected his love of nature, and he often joked that the birds helped him with his composition, even when he was completely deaf. Beethoven's music, together with that of Mozart, has been shown to stimulate parts of the brain and the calming powers of both are well known by physicians.

The later Russian composers, Tchaikovsky, Taneyev, Rachmaninoff, Borodin, Prokofiev, Glazunov, Mussorgsky, Shostakovich and others wrote wonderful music that reflected their times, including periods which were troubled. Shostakovich's famous Leningrad symphony (No. 7) was played among the rubble of the city while the Nazis were, quite literally, at the gate. The shock of hearing this must have finally persuaded them that the Russians were *never* going to let go of their city, even though hundreds of thousands of them had already died. Imagine the impact upon Russian morale of hearing this tremendous 78-min wonderful work. Prokofiev's Alexander Nevsky Suite, although ostensibly written for a propaganda film, manages to reflect the Russian spirit in many ways. It shifts from the dramatic to the beautiful and even includes Prokofiev's playful humour in places. It is interesting that many of the great Russian composers also, at one time or another, wrote music for the Russian Orthodox Church. And this music really reaches the soul. Indeed, the sound of Russian choirs, singing this music in Russian churches or cathedrals is inspiring and says much about the Russian aesthetic and spirit.

In all of these cases, the music reflects the civilisation that was in place at the time. Equally, it was an important part of the culture of which it reflected. This is something that cannot be influenced from outside. Music *always* reflects the culture that created it. It is an important release of energy and thinking. Consider, for example, the delightful Cuban music of the 1950s and early 1960s. It reflected an energy and joyfulness that would not be repressed, even though its origins were often poor. Today, so much popular music is hard edged, simplistic and ugly. This is reflecting, in many cases, a fractured culture that is wanting to be heard. However, its relative lack of sophistication is somewhat worrying. Of course, as with all art, one can only generalise to a degree. There are always exceptions, and, sometimes, these exceptions will take the art form in a completely new direction.

And so, music and civilisation are intertwined. When civilisation is flourishing, the music of the time reflects this in both its nature and sophistication. When civilisation is in decline, the music becomes fractured and uncertain. Globalisation has had an effect upon music in that popular styles have become mixed, and sometimes, it seems that the lowest common denominator prevails. The wonderful legacy of classical music remains, although the manner in which it is presented and played has changed over the years. However, classical music is a very important part of education, and all children should be exposed to it and given an understanding of its history and its beauty as an art form. It is worrying to hear adults state proudly that they 'never listen to classical music' as though it were a sin to do so. And, sadly, there are a great many who fall into this category. You would think that, as this music has been enjoyed for hundreds of years, curiosity alone would spark an interest. But for many classical music is seen as something for snobs and those who live high on the hill. And yet, classical music was often written for the masses, especially choral music, operas and other forms. Furthermore, concerts were often given in open air theatres or to commemorate special occasions. Handel's 'Water Music' would have been heard by throngs of people lining the Thames as the royal barges slowly passed by. What we now call classical music was the popular music of the day among those who understood its importance and many had access to it, regardless of their status. Over the centuries, in many ordinary households, music was practised and considered a very important part of life. Of course, some of the great composers benefited from patronage of the wealthy families, but that did not mean that their music was not heard and enjoyed by a much wider audience. It travelled readily across geographic boundaries as well. In Tudor times, John Dowland travelled widely throughout Europe, and his music would have been played in every country, both by the troubadours and within the courts of noblemen. Serious music exists to be enjoyed by everyone, and introducing a child to classical music and its history is one of the greatest gifts that a parent can give, regardless of their status. It is all so accessible now. And it is a vital part of ongoing civilisation. There has perhaps never been a time when this is more important than today, when so much is false and ugly and the individual becoming increasingly meaningless. Music can lift the spirit under such conditions, providing it is allowed to do so. It is also a vital part of history and the route to understanding that different times existed when creating that which was beautiful was seen as important. It is part of our legacy as human beings and this is very important.

Chapter 25
Where Is the Future for Serious Music Being Produced Now?

In the early days of recorded music, there was always a method of archiving the recordings. When discs were cut directly into a horn-based recording machine, a master stamper could be produced from which other discs could be made. When tape machines arrived, we had the master tapes which could be kept as a reference and from which disc stampers could be made. Sometimes, several master tapes would be produced and sent to different countries, where disc stampers could be produced from them. It was a similar situation with film, whereby multiple copies were made, in order to be shown in cinemas around the world, while a true master copy would be archived. In Britain, the British Film Institute National Archive was established in 1935, in the author's home town of Berkhamsted. It has since built purpose made premises in Warwickshire and has a centre in London.

For audio recordings, the master copies of tape recordings are stored in various locations. The larger studios often had their own archive, with recordings kept in temperature-controlled store rooms, and most countries stored recordings, like books, within a National Archive. In Britain, The British Library has an extensive archive of recorded audio, as does its counterpart in many countries. Several universities maintain an audio archive. In America, Stanford University has such an archive. And, of course, the major record labels maintain their own archives. In Britain, EMI has an extensive archive of audio recordings, dating right back to the start of recorded sound. This diversity of recorded sound archives is not necessarily a problem and may, in fact, be an advantage, as it is possible that, for some recordings, several 'master' tapes might exist, buried in archives around the world. However, the record companies themselves are the best source.

Archiving recorded tapes has an inherent problem, and that is that magnetic tapes deteriorate over time. The oxide itself can become detached and 'print through' can occur whereby a particularly strong sound might interfere with the layer of tape adjacent to it, effectively duplicating itself. All of these problems are understood, and proper tape archives would store the tapes in temperature controlled rooms and within strong packaging. It was thought that tapes had a shelf life of around

© The Author(s), under exclusive license to Springer Nature Switzerland AG 2021
J. Ashbourn, *Audio Technology, Music, and Media*,
https://doi.org/10.1007/978-3-030-62429-3_25

20–30 years, after which, they should really be duplicated. Of course, nowadays, they can be digitised and stored on computer hard drives. Storing on media such as hard discs (or even media cards) introduces other issues, as file corruption can occur and, in any case, such archives require careful management as they grow. The advantage is that hard discs can be very large and are relatively inexpensive. At the time of writing a 2 terra byte hard disc may be bought retail for less than 60 pounds sterling, and you could store a lot of music recordings on 2 terra bytes, even in high resolution. In addition, the physical space required to store thousands of recordings is considerably reduced.

Storing on hard disc and automatically producing duplicate, backup copies, is a simple matter, especially using RAID (Redundant Arrays of Independent Discs) technology. The question is what software will be used to catalogue them and access them when required. In the computer world, things change all the time. If we look back over the past 30 years, hardware has changed, operating systems have changed and applications have changed. In addition, in most cases today, software applications require helper applications or 'run-time' applications which themselves change over time. Even computer file systems change over time. This means that a computer-based archive must be constantly checked and, in some cases, completely refreshed, in order to keep up with the technology. This in itself is not so much a problem, but it does introduce the possibility of creating errors. Consequently, a computer-based sound archive must be very carefully managed.

However, there is another issue. With computer-based recording now ubiquitous and relatively easy to accomplish, the sheer number of recordings being made now is enormous, even within the serious music field, within which we may include classical music and traditional music from around the world. The major record companies will have their own digital libraries and may decide which recordings become archived. It is doubtful however that *every* recording made will be archived. And for those that are, will a time limit be applied to them? The smaller, more specialist record labels might have a particular problem in this respect. Some of them might routinely archive every single recording and some may not. What of the recordings that are made but which, for one reason or another, are not released? Will they also be routinely archived? Then there are the recordings made by independent audio engineers who may have their own archive, or may not. There are a large number of recordings being made now within the serious music field. It is hoped that all will be archived somewhere, but finding these archived masters 20 years or more on, might not be so easy.

In the popular music field, there is an enormous amount of music that is produced and placed on the Internet, without going anywhere near a conventional record company. These music files will stay on the Internet for some time, but not forever. The companies providing such services will be able to maintain copies for a while on their computer servers, but the sheer amount being produced, including as high-resolution files, will make it impractical to store everything, forever. There will also be the temptation to use cloud technology, buying disc space from a third party. This immediately introduces other issues, including around the security of data. Furthermore, if cloud storage is used, the user (i.e., the record company) will have absolutely no idea where the data is actually being stored. In all probability, it will be in another country. The only safe way of storing recordings digitally is for the record company concerned to maintain their own servers, in a secure location, and to make sure that everything is backed up (preferably in another location) and to maintain their own access software. The digital age brings advantages and issues of its own.

The primary issue here is that we have already amassed a vast quantity of recorded performances. It is important that we do retain these older recordings for a variety of reasons, including the fact that many of the performers are no longer with us, but also because of the changing style and sound of performances. One might argue that there is a limit to the number of recordings of Beethoven's symphonies that we really need or of Mozart's operas or any of the works of the great composers. However, as recording technology changes, and as younger performers and conductors enter the field, there would seem to be no reason to deny additional recordings of the classical repertoire. In the popular music field, the music itself is constantly changing, and so there would be a good case for capturing these changes which, indeed, reflect our ever changing culture.

And so, yes, we need to continue to record good performances of classical and traditional music, as well as new music in all styles. But we also need to acknowledge that the way we distribute music has changed, and consequently, we must consider how we are going to store both existing and new music for archival purposes. This should be a question of both national and global interest (Fig. 25.1). Another problem arises from the business perspective, especially when smaller record companies are absorbed into larger ones. In some cases, the output of the smaller company is simply discarded. One can think of some quite high profile artists whose back catalogue is simply not available any more because the master tapes have not been kept. If this was a problem with physical media, it will be a larger problem with virtual media and computer files. The reality is that there are no standards in this respect and what is kept or not kept depends upon the whim of the creating organisation. Even that which is kept may prove difficult to access in future years as computer technology changes and, if back catalogue is to be transformed into another format, who is going to perform this task? Some will say that this is the least of our problems at the moment, but it bears thinking about.

Fig. 25.1 The traditional way of archiving film and sound

Chapter 26
Are Advances in Technology Always Good?

Our modern world is founded upon constant change. Indeed, changes in all areas of arts, sciences and culture as civilisation itself changes. Change in consumer goods and related services drives a good deal of technology change. But if we take any one technology and examine the life cycle of changes, we usually see something of a bell curve. There is an initial slow rise, followed by a rise that is almost exponential, followed by a slowing down and a levelling off, followed ultimately by a decline. The decline is not a decline in the progress of change, but a decline in the usefulness of application. The computer industry and, in particular, the associated software industry is a very good example of this. We saw a gentle introduction, followed by rapid acceptance and a drive in specifications, followed by a lull in any useful development and then a decline as the quality of hardware has decreased and the usefulness of software also decreasing while the software itself becomes increasingly bloated, inefficient and unreliable. It has been the same story with the motor car and many other areas of technology.

In sound technology, there has been a kaleidoscope of progress curves, due to the fundamental changes in both recording and playback technology. With recording technology, we have seen major changes such as the invention of stereophonic sound, the tape recorder and now digital recording technology. With tape, we saw the number of available tracks extend to 24 and even 32, with tape widths of up to 3 in. Although, even here we saw a natural roll off at 24 tracks on 2-in. tape, as there really was not any need to go beyond this for most purposes. The machines themselves had to be ruggedly engineered in order to withstand almost constant use while maintaining good performance. We saw associated changes in the development of noise reduction systems but, otherwise, things had already levelled off and were starting to decline when digital recording arrived.

The original CD standard of 16-bit 44.1 kHz is actually quite capable of capturing good quality music. The fundamental difference in digital 'sound' saw a determination to improve on this, hence the introduction of SACD which did not attract much interest. Nevertheless, audio engineers argued that 24-bit 48 kHz was a better

J. Ashbourn, *Audio Technology, Music, and Media*,
https://doi.org/10.1007/978-3-030-62429-3_26

option, and this is what much of the film and broadcast industry use. However, there are those that wanted more and so sampling frequencies rose to 96 kHz and then to 192 kHz and beyond. The considered benefit of these higher sampling frequencies lies in a smoother and more detailed sound, particularly in the middle frequencies. However, in order to play back recordings made at higher resolution, one needs specialist equipment. Recording at 24-bit 96 kHz has been with us for some time now, but not many have embraced the format, let alone higher resolution. Public taste tends to make up its own mind, and it seems people still like their CDs. Ironically, there has been a mass acceptance of lower quality sound in the form of MP3 files, which users can play on almost anything. So, high-resolution recording is with us, but where is it going? If SACD did not catch on, then the market is surely telling us something in this respect. Still people will come up with ADCs and DACs capable of extreme sampling rates and increased bit depth. They will be of interest to a minority of audiophiles.

Audio playback has changed, from physical media on dedicated players to downloaded files placed on music streamers, connected to traditional high-fidelity systems, most of which will not be capable of dealing with high resolution. In fact, on some older systems, it might be positively harmful to put high energy, high frequencies through them as they were never designed to deal with them. On new, super expensive, so-called high-end systems, they may be capable of reproducing high resolution files, but whether the listener is really capable of hearing them is another question. Furthermore, in many cases, room anomalies will tend to introduce their own distortions which negate the benefits of high resolution. Nevertheless, there exists a small market for these high end audio components, the cost of which tends to be out of reach of most people.

The question, when it comes to audio replay, is do these systems really sound any better than our affordable analogue systems? Many think not and there is a revival in vinyl records and turntables which is quite interesting. Indeed, many system components which were considered middle range in the 1960s and 1970s are now highly sought out for their sound quality and reliable build. Even taking the natural ageing of semiconductors and other components into consideration, many of these components still sound better than modern systems and, most certainly, are better built, unless you are prepared to spend more on your audio system than the cost of your house. Loudspeaker development also seemed to follow a bell curve and some of those from the 1970s and 1980s still sound remarkable good.

The longevity of consumer goods, as we all know, has reduced considerably. The main reason being that manufacturers have committed to high turnover in both numbers of units made and revenue. Consequently, almost everything is of poorer inherent quality now than it was a few decades ago, unless you are prepared to pay astronomical prices, and even then, there is no guarantee of quality. The same is true of audio components. Technology moves forwards, but quality does not follow. Economists will argue that it is unreasonable to expect manufactured goods to be well made or to perform well, because this is not what industries are founded on. The recorded sound industry has managed to reinvent itself a few times, but has surely reached or passed its peak. The distribution of audio may continue to change, but it is hard to see further fundamental changes in how we record and distribute audio. The changes we have seen in recent years are mainly to do with poorer quality at higher prices, coupled with a good deal of propaganda. If you wish to buy a good quality audio playback system now, you will have to pay dearly for the privilege. The available recordings will be of variable quality, depending upon how they have been undertaken (Fig. 26.1). Middle range high fidelity components have all but disappeared. So it is no longer possible for the average man in the street to build a good quality system at a reasonable price. Generations have consequently grown up with poor quality sound and low expectations of replay component quality. This does not represent a step in the right direction.

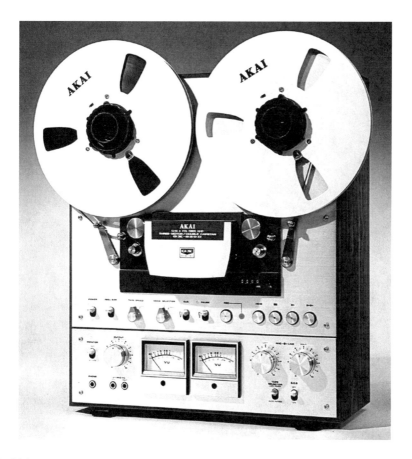

Fig. 26.1 1970s audio components were of a high build quality

Chapter 27
Teaching Audio Engineers

In the very early years of sound recording, there were no audio engineers as such, however, Alan Blumlein's invention of stereophonic sound brought scientific thinking to what had been a fairly basic technology. In particular, the placement of moving coil microphones and the higher quality tape recorders which were available in the early 1950s brought a new level of sophistication to audio recording. At first, there were 'tape men' and 'microphone men' who perfected their particular operations, but these duties quickly became the preserve of the audio engineer.

Passing these skills on would have been normal practice to companies such as EMI, who will have trained their own personnel in their own manner. Junior employees will have learned from their more experienced colleagues and would have been taught the current best practice in areas such as microphone technique and tape editing. In one respect, it was all fairly straightforward in those early days, but then, they were very much 'learning as they went', particularly with factors such as precise microphone placement and how far they could push tape recording levels.

For quite a while, record companies would simply train their own audio engineers. Of course, these individuals had to have an understanding of both physics and music and, of course, an interest, if not passion, for audio recording. As freelance audio engineers and producers came onto the scene, many of them did not have this solid grounding that experience with the major recording companies provided. Consequently, some specialist external training providers started what came to be known as 'tonmeister' (audio engineer) courses with appropriate accreditation. These days, tonmeister or equivalent courses are provided by several universities. Looking at some of these courses, it seems that some of them place more emphasis on one aspect of learning than others, with some of them heavily entwined with certain software suppliers.

A well-balanced tonmeister, or equivalent, course should cover music theory and practice, including the compulsory learning of at least one conventional musical instrument. The audio engineering side should cover an in-depth understanding of acoustics, microphone design and application, true stereophonic sound, the use of mixers and several recording media including tape, stand-alone digital recorders and more than one software application. The course should include a good deal of practical recording, both within studios, and at external live events. The resultant recordings should be carefully analysed and peer reviewed. Naturally, the use of external processors should be covered, although this should be placed in perspective. Mastering, using specialist software, should also be included and I notice that this is not listed in some courses, and yet, this is a key requirement if audio engineers are to become freelance. Nevertheless, most of these courses seem to cover a broad spectrum and typically have a duration of 3–4 years (of term time). Naturally, the purveyors of these courses like to post details of successful graduates, and here is where some doubt creeps in. The ratio between students and successful graduates seems to be quite low. It may be that recording studios, both independent and those owned by the recording companies, still like to train their own personnel. With independent studios, it may also be that they have access to a wide network of existing, successful audio engineers and producers on whom they can call. Or it may simply be that there are few opportunities that are not filled by these networks.

It may also be that these courses, while wide in scope, do not train individuals to work in the way that professional engineers and producers *do* work. There are other factors which will creep into actual projects and those in charge of the budgets will want to use people they know. In this respect, it is curious how many in the industry are of advanced years. And there is a wide gap between the knowledge and experience of these individuals and those producing 'hit' records in their bedrooms, who have no practical experience at all.

So, what would the ideal course look like? Firstly, it should take people who already have some sort of qualification in the sciences. It should then spend some time purely on physics and acoustics, including studying the practical effect of different acoustics and how to deal with them. An in-depth study of microphone design and why some microphones from the 1950s and 1960s are still favoured, including the fundamental types and a look at Ambisonics. Practical recording with minimal equipment should factor strongly, as should an understanding of what stereophonic sound actually is. Time spent in more than one commercial studio should follow, before looking at the very best software (not the most popular) and how it differs from lesser offerings. Practical field work should follow recording live events across a broad range of styles, the results of which should be carefully analysed. It should not take 4 years to study such a course, but a continual study of 2 years should get the student ready for practical experience in which, naturally, he or she would continue to learn.

One problem within the field of popular music is that many so-called hits (although it is hard to measure this these days) are performed and produced by individuals with no experience at all. Those commercial studios who work predominantly within this field will have their own way of working which, in many cases, will certainly have little to do with good audio engineering but much to do with the use of popular software applications. This leaves the classical and traditional music fields and, even here, we find commercial releases that are certainly not true to the original sound, especially, oddly enough, from the very big recording labels. Proper audio engineering and mastering seems to being pushed towards the smaller, specialist labels who might also use existing freelance audio engineers whom they trust. Consequently, the route for someone who is passionately interested in high-quality audio engineering is a confused one. A successfully undertaken tonmeister course would certainly do no harm, although it is by no means a guarantee of success. Getting to know the specialist labels and how they operate would also do no harm, and in this respect, learning how to put yourself across in a professional manner will be important.

Of course, some will be more interested in audio for film, multimedia and other more general areas within the performing arts. There are residential courses which focus on such things, some of which are very good. The other way of gaining experience is simply to volunteer. For everything, whether amateur or professional and, by so doing, become involved in audio within your own geographic area. It is often surprising just how much goes on from a local and regional perspective. Things are very different now from how they were in the 1950s and 1960s. One needs to understand this and consider what part, if any, of the modern scene one wishes to be associated with. The really brave will simply go their own way and make their own recordings, building their own network as they go. If they are good, this will become apparent, sooner or later. However, to do this, they *must* have a solid grounding in physics, acoustics and music, as well as an understanding of modern equipment. And, of course, the best way of learning is via practical experience. Regional amateur orchestras may offer opportunities in this respect. In any event, training to be an audio engineer within the current climate is a confusing business. The problem mainly being that standards are simply not being maintained even within the industry, so what, exactly, is the trainee training to do that is going to be considered useful? One would like to see a program that focuses little on software and very largely upon audio quality and how to obtain it. Such programs however seem conspicuous by their absence. The ones which do exist may be trying to cover too much ground and, perhaps, we need to think more about specialisation.

Chapter 28
The Future: Here It Is

The future for recorded sound is looking a little confused at present. The big record labels are continuing to produce CDs, but the manner in which they are being purchased is changing. There used to be dedicated record stores, selling mainly CDs, but also cassettes and vinyl records. They have all but disappeared as more and more people buy online. This is a great shame as the opportunity to browse and find interesting items that you had not thought of has simply gone. Most high streets would have two or more dedicated record stores with knowledgeable staff. Now there are none. Most of the big newsagent chains also used to sell CDs. Now they do not. And so consumers, whether they like it or not, are forced to buy online. Usually at high prices for current recordings.

However, more and more people that the author comes across are saying that they no longer have a CD player in the house. These are usually the same people that do not have books in the house. So, how do they listen to music? Via MP3 files downloaded from the Internet or on YouTube. Which means that they listen to short bursts of music, but nothing more. This is alarming for many reasons, but especially because it is causing them to have short attention spans. This seems especially prevalent among younger generations. Of course, not everybody is like this, but there has definitely been a shift in this direction culturally. The author is thinking of two of his acquaintances. One of them has a small audio system and a handful of cheap compilation CDs, the other has no audio system at all. Both of these individuals have no books in the house (except one or two that I have given to them) but have huge, very expensive televisions. They both claim that they do not like classical music and that they do not have the time to read anything, although both are retired and wealthy. If this indicates a general trend, then it is particularly worrying as it means that generations are making their way through life without enjoying the wonder of the arts. Again, there are, thank goodness, plenty of exceptions. Those who are in a privileged position and attend the best private schools will be exposed to both music and literature, and it is from among these people that we shall find our future musicians and writers (and painters, although this is a different story). These

J. Ashbourn, *Audio Technology, Music, and Media*,
https://doi.org/10.1007/978-3-030-62429-3_28

individuals will experience the wonder of real music, but how will they purchase it as consumers?

There are an increasing number of Internet sites that provide music of all kinds as downloadable files. Some of them provide these files in high resolution, but in order to be able to play them, you need a digital audio player of some kind. This is the way that many young people enjoy music. However, not many young people have been exposed to really good recordings, played on good quality equipment, and so they really have little appreciation of quality. As most of them have mobile phones, they simply download MP3 files and play them on their phones, listening with low-quality 'ear buds' as they have become known. It is unlikely that they will listen to a complete symphony in this way, and so, for many, their experience of music will be limited to relatively short bursts or popular music singles. Some of them will, of course, buy better quality digital audio players, but there are so many available at relatively low cost that they may still not hear music as it should be heard. In order to hear good quality, one must pay for good quality equipment, whether it be portable or designed for use in the home.

It is clear then that the way younger generations are acquiring and listening to music is different. There is a swing towards downloading music files from the Internet, sometimes directly to mobile phones. It also seems to be rare that individuals sit down and really listen to complete classical works. Those that do will be exposed to a wonderful world of music and inspiration. Those that do not will miss this entirely. It is the same with classical literature. Dedicated book shops are becoming fewer, and we are left with one or two chains who sell mainly popular paperback romances and the like. The author remembers strolling up and down Charing Cross Road in London, where there were several wonderful book shops, selling beautifully bound copies of all the classics and who were staffed with knowledgeable individuals who would always find what you were looking for, or recommend something that you had not thought of. They have all gone now, except for one, and that one has changed dramatically.

Getting back to the future for consumed music, several factors are clear. Firstly, most music is purchased from online sources, either as CDs or as downloadable files. Consequently, the record companies tend to offer both, at least for the moment. It is also clear that younger generations have little interest in high fidelity and do not own, or intend to own, high-fidelity systems. Even if they wanted to, the middle range of good quality high-fidelity components has mostly disappeared. We are left with awful sounding cheap all in one systems and frightfully expensive 'high end' systems which few could afford. The high street electrical chains sell mostly the former, with one or two more expensive designs which are poorly made and sound dreadful. So, in terms of audio equipment, the dedicated audiophile must search far and wide to find anything of good quality at affordable prices.

The record companies will, naturally, follow the retail trends. At the moment there remain sufficient numbers of people buying CDs to keep this a viable option. But will this continue? Will we be left, in the end, with little option but to download the music of our choice from the Internet? I hope not, because the tactile experience of loading a CD or placing a vinyl record onto a turntable, is part of the enjoyment of music. Having the object safely in your possession, complete with sleeve notes appertaining to the performance, is an integral part of the pleasure of listening to music. Those, like the author, with large music collections will enjoy the pleasure of selecting a piece which has not been heard for a while and reliving the pleasure of listening to the performance. If music is reduced to MP3 files on mobile phones, then this pleasure is not experienced.

Maybe other formats will come along and catch the public imagination. Who knows? But the current situation is rather confused, especially for classical music. If orchestras are to survive, then people need to attend concerts and buy the recorded performances. At present, sufficient numbers of them do, but ticket prices are sometimes high, and the ability to replay music in the home properly is currently rather strained. There will, hopefully, be sufficient numbers of young people who will be taught music properly in schools and colleges and who will retain this interest. From their ranks come not only the consumers of music, but the performers and the audio engineers and producers. And so, this work is a plea for proper music education in schools of all types, not just the private schools. Also a plea for audio engineers to revisit true stereophonic sound and to start making stereo recordings where orchestras sound like orchestras.

Music is an integral part of our lives. Without it, we have not lived a fulfilled life on Earth. It has the ability to calm, to heal and to inspire. Furthermore, it has the ability to reach and develop parts of our intellect which are important, for us personally and for humanity in general. Everyone should experience serious music and understand its history and the lives of those who created it. Our rich legacy of music is a world full of wonder and wonderful surprises which is there, just waiting for us. To live a life and not experience it is to not understand the better side of humanity. It is the same with literature and art.

The future of music is with us now. It is polarised into the difficulty of maintaining large orchestras and recording facilities capable of housing them at one end, and the rise of instant music made possible by computers at the other. In the middle are the distribution channels which themselves are fractured into tangible product (CDs and vinyl) and online files of variable quality. The latter are becoming more prominent, but there may still be a bright future for physical product, albeit rather different from what we have known.

I am cheered by remembering Beethoven's comments to his patron Prince Karl von Lichnowsky when they fell out because Beethoven refused to play for French soldiers being entertained by the prince. He wrote a note to the prince, saying, 'Prince, what you are, you are through chance and birth; what I am, I am through my own labour. There are many princes and there will continue to be thousands more, but there is only one Beethoven'. Exactly, and that is why his music is so important. Beethoven understood that people would be enjoying his music for hundreds of years. Let us hope that they continue to do so (and in stereo). How else will they understand the man and his times? And so it is with all the classical composers. They are part of our history. A valuable legacy which we must ensure gets carried forward to future generations and in the best possible quality. If this fails to happen, for a sizeable majority, then we shall be denying them their own history, the history of civilisation and humanity. This is why serious music is so important. Let us look to the future with this in mind. Currently, we are not doing so and music is being reduced to short bursts on TV and the Internet. We need to get a proper recording, production and distribution system back in place, which enables people to see and buy recorded music in a sensible manner. However, no-one seems particularly interested in doing so. How different things have become from those pioneering days of the 50s.

Epilogue

Like most children, my introduction to music of any kind was very occasional, if my parents happened to be listening to the radio. But they never listened to classical music. My poor standard of education in a forgotten rural area of post-war, impoverished England, did not even have music on the agenda, so I had no opportunity to learn how to play an instrument or have any introduction to music theory (both were subsequently corrected by my own efforts). However, when I broke away from home and went to live and work in London, it occurred to me, first with literature and then with music, that if these 'classics' had remained popular for so long, then there must be something to them. And so I started, by saving what I could from my low wages, to buy paperback versions of the classics, which I immediately found fascinating, especially those stretching back to Aristotle and Plato. I had ready access to the National Portrait Gallery and the British Museum, both of which I visited whenever I could, and this helped me to start placing things in context (it was also the start of my enduring fascination with Egyptology).

I had built for myself a very basic disc replay system, and the first classical recording I bought was a selection of works by Handel. For me, in those days, records were expensive and I could not afford to buy many. However, I also had a radio, and in those days, early FM broadcasting in England was of a high quality, and I listened a good deal to the then good output of the primary BBC channels. I think the second record I bought was of works by Delius, which I liked but found a little too tame for a 21-year old. And then, I started to hear works by Beethoven, and that was that, it suddenly all fell into place.

Today, whenever I am feeling tired, or a little disenchanted with the world, I always return to Beethoven. There are a number of wonderful composers, stretching from Handel and Bach to Rachmaninoff and Prokofiev, and then there is Beethoven. He just seems to occupy a different space in my mind and heart. I now possess a vast library of classical music and also, to a slightly lesser degree, jazz and other forms of traditional music from around the world, but Beethoven remains supreme in my estimation. For those interested in learning more about his fascinating life, there are

© The Author(s), under exclusive license to Springer Nature Switzerland AG 2021
J. Ashbourn, *Audio Technology, Music, and Media*,
https://doi.org/10.1007/978-3-030-62429-3

several good books; however, the one I would recommend is the vast undertaking of the work by Thayer, within which are many of Beethoven's own letters that reveal much about the man.

Therefore, for those readers who may not have listened to much classical music, I recommend that you start with Beethoven. The beautiful and uplifting violin concerto is a nice, comfortable piece to whet your appetite. There are at least two recordings of it by Wolfgang Schneiderhan, whose sweetness of tone and impeccable playing are well suited to the piece. The piano concertos, numbers three and five are a logical next step which will reveal more about the composer. The EMI recordings made by Solomon are recommended as Solomon's dogged adherence to the score produce performances which are probably as close to hearing how Beethoven may have played them, or certainly intended them to be played, as we shall ever get. The often neglected Septet is a beautiful work, sometimes adapted for a different instrumental line up, but always interesting. The great symphonies, yes of course and there are several good sets available, although you might which to research and choose individual performances which have stood the test of time. Start with the third (Eroica), the sixth (Pastoral) and the ninth (Choral), and you will not be disappointed. Listen also to the Mass in C and the extraordinary but magnificent Missa Solemnis, which will leave you wanting more. By now, you will be getting a feel for Beethoven's beautiful, magnificent, dynamic and, sometimes, even playful scoring. Now move to the late string quartets and, if you have developed the ability of critical listening, let them absorb you into a new and wondrous world which is inhabited solely by Beethoven.

Mozart, yes, of course, and my preference is for the music written for the wind instruments. Perhaps it is because these were often pieces written for friends, and it is here that I think we hear the true Mozart. The clarinet concerto needs no introduction, but there is also the clarinet quintet, the various pieces for flute, including the flute concerto, the beautiful and touching bassoon concerto and more. The Mozart operas are also magnificent in their own way, especially for those with an understanding of Mozart's life. There is much to be found within them.

Let us not forget Haydn, the father of the symphony. Within Haydn's enormous output, there are many gems to be found. Often, listeners steer towards the later works, especially the 'London/Oxford' and 'Paris' symphonies, but there is much to enjoy within the early works as well. Haydn's music is often deceptive. It has inherent tunefulness hiding a sophisticated underpinning of imaginative and sometimes complex scoring.

Of the many that came after this high point in Viennese culture, including Schumann, Schubert and many others, I focus upon two, in particular, with which to broaden the appetite of the new listener. Firstly, Brahms. His output may not have been quite so prolific as Mozart's or Beethoven's, but it is always interesting. His symphonies manage to say something new and different, and there are several offshoots into areas that interested him. And, of course, there is the wonderful Alto Rhapsody. There are, it seems, just a handful of modern recordings of this piece. If you seek out the 1947 recording, conducted by Bruno Walter and featuring the uniquely beautiful voice of Kathleen Ferrier, you will perhaps understand why. You should really get both, the 1947 version and any one of the very good more recent recordings. The early recording has a palpable tension, released into an opening up of beauty that is almost shocking. No-one will ever replicate Ferrier's control of this. However, the later recordings are mostly worthwhile.

It is no secret that Wagner is an acquired taste. His majestic orchestral scoring, coupled to a penchant for a heroic and sometimes tragic opera, makes some of his works, like The Ring, a challenge for some. However, if you allow yourself to be drawn into Wagner's world, it all suddenly becomes clear. My personal favourite is Tristan und Isolde, a tragic but beautiful love story conveyed by equally moving music and words. It would make a nice entry into Wagner's world.

The Russian composers have left us a particularly rich legacy of more recent times, but within the classical tradition. Tchaikovsky gave us some wonderful tunes, especially from his ballets. These are, perhaps, what most are familiar with, but there is also the beautiful and intense violin concerto which reflects Tchaikovsky's personal struggles at the time. There are also the symphonies which are rich, but, perhaps in a slightly understated or, at least, different manner. There is also a rich vein of compositions for piano which are seldom heard. The Russian pianist Viktoria Postnikova has recorded a delightful set of them. The 1812 overture needs no introduction, and there are more recordings of it than you can shake a stick at. Look at those recorded in the late 1960s, among which, you will surely find some gems. Tchaikovsky, like many of his contemporaries, also wrote music for the Russian Orthodox Church. Seek it out. It is wonderful.

Shostakovich is a composer whose work I, probably like many non-Russians, found a little difficult at first. But this is music written often from the heart and soul, like his wonderful seventh symphony, sometimes called the 'Leningrad'. This is music with a sense of purpose and which demands attentive listening. Indeed, most of his work falls into this category. The symphonies (all of them) and the string quartets are of particular interest and will take you into another world.

Rachmaninoff of course left us the wonderful piano concertos, full of drama and passion, but he also wrote for the Russian Orthodox Church, including his 'Vespers', and Liturgy of St John Chrysostom. These are beautiful works.

Taneyev, Borodin, Glazunov, Khachaturian, Mussorgsky and many others, all left us with some remarkable works. But then there was Prokofiev, my favourite in many ways. Some will find Prokofiev's work a little difficult at first as it is different in several ways, but this is simply because, as a composer, Prokofiev seemed to be able to turn his hand to anything, although his playful sense of humour often breaks through. Who else could write an opera entitled 'The love of Three Oranges'? (which, by the way, is a wonderful work, especially if you see it performed by a Russian ballet troupe). The listener might also be surprised at how many little tunes are already familiar to them, not knowing who the composer was. A work of particular interest is the Alexander Nevsky Suite. Prokofiev was 'asked' to write this piece which was originally to accompany a Russian propaganda film, and yet, in addition to the expected traditional themes, he manages to inject a good degree of satire and unexpectedly of pure beauty within the choral theme which starts very softly and repeats at intervals in a more dramatic manner. Prokofiev's symphonies are all interesting, as are his works for piano.

A note here about Russian Orthodox music, which has been mentioned once or twice. The Russian Orthodox religion is something taken very seriously. Several attempts have been made to destroy it, from Ivan the Terrible through to Stalin and many thousands of good Russians died (i.e. were murdered) before they would renounce their religion. The music for the Russian Orthodox Church, contributed by many of the later Russian composers, is extraordinarily beautiful, especially when heard and sung by Russian choirs. There is something of the Russian spirit which is expressed within this music. My introduction to this music was when I had the opportunity to see a small Russian choir who were on a cultural visit to England. I had to drive around 40 miles to a not very inspiring modern church in Oxford. The concert they gave was not well attended, and my wife and I sat right in the front. Within about 10 s of them starting to sing, I was absolutely hooked. They brought tears to my eyes, and I well remember my subsequent conversation with them and the discovery that they were just ordinary people from a small town in northern Russia. Later, I had an opportunity to visit St. Petersburg where I saw not only another larger Russian choir but also the authentic Cossacks, which was an experience I will never forget. There is much to discover within the music and culture of this huge land.

And then, after all, I return, as usual, to Beethoven. He makes me smile, cry and simply wonder how a single individual could produce so much music of such consistently high quality. There may be much that we do not have a record of as Beethoven was his own most stern critic, as we can see from his notebooks. But what we do have is wonderful.

The value of this music, to civilisation itself, is incalculable. Consequently, it is our duty to perform it accurately, as the composer intended, and also to record it properly, in a way which captures the sound of an orchestra in performance. Alan Blumlein showed us how we could do this in 1931. Returning to these first principles of stereophonic sound would serve us well. And this is partly why this book has been written. I do hope that you have enjoyed reading it. I also hope that it might set you off on your own musical journey, whether it be performing, recording or simply listening.

Julian Ashbourn, September 2020.

Index

Printed in the United States
by Baker & Taylor Publisher Services